The Colorado Labor Wars

Cripple Creek 1903–1904

Regional History Series
Editorial Committee
Tim Blevins
Chris Nicholl
Calvin P. Otto

Cover and Book Design
Nancy Thaler

Acknowledgments

We are grateful to the many people who made the symposium and this book possible. We especially want to thank Elizabeth "Betsy" Jameson, Larimore Nicholl and Katherine Scott Sturdevant for their editorial assistance. We wish to also thank Betsy for proposing the symposium commemorating the Cripple Creek Labor Wars, resulting in this publication. The Editorial Committee extends its appreciation to the Pikes Peak Library District's Special Collections staff who all worked hard to make this event possible. A special thanks to Dee Vazquez, Community Relations Officer, her staff, and Steve Antonuccio, Multi-media Center and Production Studio Manager, for video recording these lectures.

The Editorial Committee

The Colorado Labor Wars

Cripple Creek 1903–1904

A Centennial Commemoration

Edited by

Tim Blevins, Chris Nicholl & Calvin P. Otto

Published by

ppld

PIKES PEAK LIBRARY DISTRICT

The Colorado Labor Wars: Cripple Creek 1903–1904
A Centennial Commemoration

Third printing.

For additional copies or DVDs contact:
Pikes Peak Library District
P.O. Box 1579
Colorado Springs, CO 80901

Published 2006
ISBN-13: 978-1-56735-225-2
ISBN-10: 1-56735-225-1
Library of Congress Control Number: 2006901084
Printed in the United States

This publication was made possible by private funds.

CONTENTS

Elizabeth Jameson • 3
History, Memory, and Commemoration:
The Cripple Creek Strike Remembered

Marcia Tremmel Goldstein • 35
"Meet Me at the Ballot Box": Women Voters, Striking Miners,
and Colorado Labor Politics, 1893–1904

Bridget Burke • 53
Is Colorado In America?: Emma Langdon, Polly Pry,
and the Colorado Labor Wars

Jan MacKell • 71
Prostitution During the Cripple Creek District
Labor Wars, 1904

Ed Hunter • 77
High-Grading: Theft or Wage Supplement Right?

Katherine Scott Sturdevant • 83
John Harper, Deported Miner:
The Cripple Creek Strike as Family History

Elizabeth Jameson • 101
Talking the Walk: The 1903-1904 Strike Centennial
in the Cripple Creek District

Selected Bibliography • 117

Symposium Contributors • 121

Index • 125

Pikes Peak Regional History Symposium
Sponsored by
Special Collections, Pikes Peak Library District
Pikes Peak Regional Labor Council

In partnership with
Colorado Springs Pioneers Museum
Cripple Creek District Museum
Western Museum of Mining and Industry

Project Director
Chris Nicholl

Pikes Peak
Regional History
Symposium Committee
Chris Nicholl, Co-Chair
Calvin P. Otto, Co-Chair
Steve Antonuccio
Tim Blevins
Ed Hunter
Linda Lemieux
Jan MacKell
Matt Mayberry

FOREWORD

The purpose of the first Pikes Peak Regional History Symposium was to commemorate the centennial of the Cripple Creek Labor Wars (1903–1904) and to detail the causes and consequences of one of the era's violent labor strikes that spread throughout the Colorado mine fields. Secondarily, our purpose was to publicize a portrait collection, housed in the Pikes Peak Library District's Special Collections, of ninety Western Federation of Miners members who were arrested and held by the Colorado State Militia; many were deported from the state as a consequence of their union affiliation.

The free public symposium, held on June 5, 2004, was a collaborative effort involving Pikes Peak Library District's Special Collections, Colorado Springs Pioneers Museum, Cripple Creek District Museum, and Western Museum of Mining and Industry, with financial support from the Pikes Peak Regional Labor Council. The daylong program took place in the 1905 Carnegie at Penrose Library. The event was completed the next day with an early-morning vigil and an afternoon walking tour of the major strike sites in Victor, led by Dr. Elizabeth Jameson.

This publication contains the symposium papers presented by the authors and is intended to contribute to the scholarship regarding the events that occurred in the Cripple Creek District during the Colorado Labor Wars. We are appreciative of everyone who contributed their research to the symposium and this publication, as well as to those who attended the events. The program was videotaped for screening on Adelphia Channel 17 — The Library Channel, and DVDs are available for purchase from the PPLD Multi-media Center and Production Studio.

Paula J. Miller

Executive Director

Pikes Peak Library District

Memory: The power, act, or process of remembering. The length of time over which remembering extends. Fame after death; posthumous reputation.

Remember: To have (an event, thing, person, etc.) come to mind again; think of again. To bring back to mind by an effort. To keep in mind with some feeling, as of pleasure, gratitude, etc. Remember implies a putting oneself in mind of something, often suggesting that the thing is kept alive in the memory so it can be called to conscious thought without effort.

Commemorate: To honor the memory of by a ceremony, etc. To preserve the memory of; serve as a memorial to.

Dismember: To cut or pull to pieces; separate into parts. To remove the limbs from.

History: The account of what has happened. Narrative, story, tale. All recorded events of the past. The branch of knowledge that deals systematically with the past; a recording, analyzing, co-ordinating, and explaining of past events. A known or recorded past.[1]

History, Memory, and Commemoration: The Cripple Creek Strike Remembered

Elizabeth Jameson

This essay was prepared to commemorate the centennial of the Cripple Creek strike, one of a series of strikes known as the Colorado Labor Wars of 1903–1904. A weekend commemoration organized by the Pikes Peak Library District, exactly one hundred years after the strike's dramatic climax, included a symposium on June 5, 2004; a 2 A.M. vigil June 6, 2004, below the former Independence, Colorado, townsite; and a tour of the strike scene in Victor, Colorado, on June 6. These events linked the sites most directly involved in the conflict, bridging the divisions of interests and ideologies that fueled it and emphasizing the different meanings the Labor Wars carried for the participants and their descendants.

Commemoration implies a shared understanding of events, a shared set of symbols, a common cast to memorialize. Histories of the Colorado Labor Wars establish no such common ground.[2] Selective memories and erasures recorded different versions of this history.

Open a history book and look for the Cripple Creek District. Most U.S. labor histories mention Cripple Creek, and the district has written romantic and bawdy chapters in local Colorado lore. The cast of characters includes the romantic failure: Bob Womack, a hard-drinking ranchhand who spent twelve years prospecting before finally, in 1890, striking paydirt in a secluded cow pasture, who sold his claim for $500 and a case of whiskey because he couldn't afford to develop it. It includes the lucky winners: Winfield Scott Stratton, James Burns and Jimmy Doyle, the trio of working guys who struck it rich on Battle Mountain. And those who won in the long run: Charles Tutt, Spencer Penrose, Charles MacNeill, and Albert Carlton, memorialized in local histories as the lovable roguish youngsters who built their empires by day and played equally hard at all the amusements Cripple Creek's bawdy Myers Avenue offered young men with gambling spirits. And the women who worked the vice district: Hazel Vernon, Belle McCluskey, Leo the Lion, romantic, colorful, one-dimensional caricatures of real women who earned their livings in the local sex trade.[3]

Open a labor history and there they are again, locked in an equally romantic conflict with Big Bill Haywood and the Western Federation of Miners, the forces of law and order battling a faceless band of miners blindly following their socialist leadership. So add to the cast Big Bill, taking on the capitalists, sending his troops into a losing battle. And the Colorado governors, separated by a decade and a class divide. Gov. Davis "Bloody Bridles" Waite, aging Aspen Populist, swung the power of the state behind the miners in the first Cripple Creek strike, in 1894, and won for them the right to belong to a union, an eight-hour day, and a $3 minimum daily wage. Gov. James Hamilton Peabody, Cañon City banker, sent the militia back in 1903–1904, led by local mining superintendent Sherman Bell who announced his mission to "do up this damned anarchistic federation."[4]

These colorful characters told the story of Cripple Creek for generations. But they aren't the people I first remembered when I thought about what we commemorated, whose stories I honored, who remain the reason the rest of it matters. For me, the centennial evoked memories of Leslie and Ruby Wilkinson, who built the road to the top of Mt. Pisgah outside Cripple Creek; of Beulah Pryor, who came to the district during the 1903–1904 strike when her stepfather was recruited as a strikebreaker; of Charlie Gatheridge, who came in 1905, just after the district became a non-union camp. I thought of Kathleen Welch Chapman whose parents were union activists in both Cripple Creek strikes, and whose family story encapsulates the union side of the history.[5] They were all in their 80s and 90s in the 1970s when they shared their memories with a very green young historian with a tape recorder who asked impertinent questions. Most of all I thought of May McConaghy Wing, born in Leadville in 1890, the daughter, wife, and mother of hardrock miners, the keeper of local memories who ran the Victor Museum from the time she was 70 until she was 90.[6]

May Wing voiced the challenge of writing this history. "I lived the history that I can tell," she told me. "And of course, the history today in books that's written a lot, is not really the true thing, as it was lived."[7] The imperfect connection between history "in books that's written a lot," and history as it was lived, lay at the heart of the strike commemoration.

For May Wing and countless others, Cripple Creek was more than the name of a dramatic strike. It was home. Known as "the World's Greatest Gold Camp," by 1900 the district's ten towns housed some 32,000 people, most of them miners and their families drawn by the promise of steady work at good pay. The district was, from 1894–1904, a stronghold of the Western Federation of Miners (WFM), renowned for its strikes, its endorsements of the Socialist Party, and its leadership in founding the Industrial Workers of the World (IWW). Founded in 1893, the WFM's first major victory was the Cripple Creek strike of 1894. The victorious miners supported other crafts, and by 1902 helped organize a majority of all workers in all trades in some fifty-four local unions.[8]

View of Anaconda, Colorado, with houses, tailings and the Mary McKinney and the Morning Glory mills on the hillside. *Photograph by L. C. McCLure, from Special Collections, Pikes Peak Library District.*

The unions influenced wages, hours, and working conditions, provided for the health and welfare of their members, established libraries and reading rooms, published a daily newspaper, the *Victor and Cripple Creek Daily Press*, which boasted that it was "The Only Daily Newspaper Owned By

Organized Workingmen." They organized the local holidays, and wielded considerable political influence. Labor's power ended with "the" Cripple Creek strike of 1903–1904, when, for fifteen months, miners struck to support smelter workers who were fired and blacklisted for joining the union.

Much of that time the district was under military occupation. The strike ended with a tragic explosion at the Independence railroad depot that killed thirteen non-union miners, the forcible takeover of the local government, the destruction of local union halls by an employers' mob, the deportation of over two hundred union men to Kansas and New Mexico, the imposition of an owners' hiring card, and a ban on organized labor.

The strikes wrote major chapters in the "heritage of conflict" in western mining.[9] But nowhere in these histories would we find May Wing or thousands of others who built the unions and stayed to live the consequences when they lost them. How do we commemorate *their* history?

What does it mean to commemorate a strike, those who waged it, those who inherited its legacy?

Consider the words. *History. Memory. Remember. Dismember. Commemorate.* Check the dictionary. *History*: A known or recorded past. *Memory*: The power, act, or process of remembering. *Dismember*: To cut or pull to pieces. To remove the limbs from. *Remember*: To bring back to mind by an effort. *Commemorate*: To honor the memory of; To preserve the memory of; serve as a memorial to.[10] Key words connect history, memory, and commemoration: *Power. Act. Honor. Known. Recorded.*

The history of the Cripple Creek strike is more than the events themselves. Many participants' stories endured in a submerged history of silenced stories, dismembered as surely as the figures on the Ludlow Memorial statue in May 2003, when a vandal literally ripped off their head and limbs.[11]

The events of the strike, what it meant, which of the participants should be honored or blamed, these are still, a century later, hotly contested. I don't have a definitive agreed-upon version that can end a century of debate about the strike and its meaning. I can contrast a bit of the recorded past with a remembered past of what the union's successes and its defeat meant to working miners and their families. Rather than detailing and quantifying a decade of thousands of workers' lives, I'll focus here on one miner who could represent them, and on his extended family.[12]

His name was John Welch. Born in 1864, the son of a Vermont coal miner, he entered the coal mines after his father died in a mining accident on July 4, 1875—hardly an Independence Day for the eleven-year-old who quit school to take his father's place underground. Some time in the 1880s, John Welch moved to Leadville, like many who sought the higher wages and lesser dangers of western hardrock mines. There he met and married Hannah Doran; in 1893 they moved to the new gold camp.[13] Their daughter Kathleen dated the first strike in family time, with her older brother's birth. "In 1894 Tom was born," she said, "and that's when the Bull Hill strike was." She told of how her parents joined in that struggle, even as sheriff's deputies from Colorado Springs, the mine owners' stronghold, charged the union camp atop Bull Hill. "There wasn't too many people up there, you know—and the men went out and they had their wives come along with them. And Mama said she wrapped Tom in a blanket and went along." John was, she said, "a great union man. He always believed in that union—for people and among people."[14]

John Welch risked that union in the second strike, and the vibrant working-class culture it anchored. The strike was not a simple conflict between miners and mine owners. Competing groups of owners had integrated control of their mines, mills, smelters, and railroads. The WFM was committed to industrial unionism: to organizing all the workers in the industry, miners, millmen, and smelter workers, into a single union, to match the industrial power of mine owners who competed among themselves to control the entire industry from the unmined rock to the refined gold.[15]

In August 1902, the WFM organized Colorado City Mill and Smeltermen's Union No. 125, whose members worked in the Standard, Portland, and Telluride Mills. Charles Tutt, Spencer Penrose, and Charles MacNeill, who controlled the Standard Mill, were anti-union; the Portland's President, James Burns, did not discriminate against union labor. His Portland Mine remained open throughout most of the strike, as did other union properties that did not supply non-union mills.

The Standard hired a Pinkerton detective to infiltrate the union, and then fired forty-two union men he identified. The WFM requested their reinstatement, a $2.25 daily minimum wage, and the right to union membership. MacNeill refused to negotiate; the millmen struck his mills. El Paso County deputy sheriffs provided security at the owners' expense; the sheriff deputized MacNeill for the duration of the strike. Then the governor named MacNeill and Spencer Penrose aides-de-camp on his personal staff while the National Guard was "protecting" their mill.[16]

Between August 1902, when the smeltermen organized, and February 1903, when MacNeill refused to bargain, Coloradoans went to the polls to decide a pivotal and hotly contested election. Democrats and Populists won a combined majority of the votes, but Republican James Hamilton Peabody took the governor's office by a plurality. During his two-year term, major strikes rocked the northern and southern Colorado coal fields, the mills and smelters of Denver and Colorado City, and the hardrock communities of Clear Creek County, Telluride, and the Cripple Creek District. Over the protests of local authorities,

Corporal Frederick Bridgeman Bryant (left) of the Colorado National Guard at the Goldfield camp near Cripple Creek, Colorado, 1903. *Photograph courtesy of Elizabeth Ogden, from Special Collections, Pikes Peak Library District.*

Peabody dispatched the National Guard to strikes in Telluride, the southern coal fields, Colorado City, and the Cripple Creek District, but not to Idaho Springs, where businessmen deported fourteen striking union miners from Clear Creek County.[17]

The strike spread from Colorado City to Cripple Creek when the miners refused to work on mines that shipped ore to the unfair mills, hoping that by drying up the ore supply they could protect the mill workers who labored twelve hours a day for $1.80.

Over the objections of local authorities, Peabody sent the National Guard to the district for much of the strike, at a cost of over $400,000, which the mine owners advanced in return for certificates of indebtedness from the state at 4 percent interest. The local anti-union Citizens' Alliance was mustered in to help as Company L in the National Guard, led by Adjutant General Sherman Bell, who took the job only after mine owners agreed to make up the difference between his $1,800 annual salary as adjutant general and the $5,000 he had earned managing the Moffat mines in the Cripple Creek District.[18]

Periodic violence – some major, some minor, some staged – provided the justification for the armed force. Thomas Stewart, a carpenter erecting a fence around a non-union mine, was beaten. Someone took spikes from the tracks of the Florence & Cripple Creek railroad; the trackwalker was tipped off and the incident passed without mishap.

On November 21, 1903, an explosion on the sixth level of the non-union Vindicator Mine killed Charles McCormick and Melvin Beck. The Vindicator lay within the perimeter of the military encampment, and was well guarded at the time. On January 26, 1904, faulty machinery severed the cable hauling miners to the surface of Stratton's Independence Mine; one man was thrown clear and fifteen plunged to their deaths. All sides pointed fingers, alternately blaming union miners, Mine Owners' Association (MOA) hirelings, or incompetent non-union labor.[19]

Labor also protested the excesses of the militia, which arrested union leaders and held them without charge, suspended the right of *habeas corpus*, and denied freedom of assembly, freedom of the press, and the right to bear arms.

Hoist engineer next to cable similar to the one that was severed on Stratton's Independence, causing the cage to fall to the bottom of the shaft, killing fifteen miners. *Photograph from Special Collections, Pikes Peak Library District.*

Bell's lieutenant, Thomas McClelland announced: "To hell with the constitution, we aren't going by the constitution."[20]

The strike ended tragically on June 6, 1904, when a bomb exploded beneath the platform of the Independence, Colorado, train depot, killing thirteen non-union miners coming off shift from Albert Carlton's Findley Mine. Who set the dynamite and who paid them remain debated.[21] What followed is clear. Within hours, the Citizens' Alliance and mine owners forced elected officials to resign—they told the sheriff he could resign or be hanged. They seized control of the local government, provoked a riot, smashed the union halls and the WFM's four cooperative stores, and banned union labor from the district. The militia deported over two hundred union leaders to Kansas and New Mexico; hundreds of others left in fear, or were "whitecapped" (violently forced to leave, usually by hooded or masked men) out of the area.[22]

During the strike many mines that hired union miners were forced to fire them. On June 9, 1904, the militia occupied the largest remaining union property, James Burns' Portland Mine. Accusing Burns of "employing and harboring large numbers of dangerous, lawless men," Bell announced, "I anticipate Mr.

Burns will be permanently deposed, and I hope obliterated from that vicinity." On June 19, an anti-union faction deposed Burns as president of the company he had founded, and the mine he had discovered.[23] In November, two MOA gun thugs intimidated people from voting by killing two union poll watchers, one of whom was a key witness to the investigation of the Independence Depot explosion.[24]

A National Guardsman holds the end of a rope that is knotted around a man's neck. Identified as "Lynching, Military Strikes—Cripple Creek 1903—04" fits written accounts of Sheriff Henry Robertson's meeting with mine owners on June 6, 1904, in all respects except that the rope is actually around the victim's neck. The room was on the second floor of the Victor Armory (now the Victor Elks Lodge). See page 110 this volume. *Photograph courtesy Colorado Historical Society, CHS.X4604.*

Over a year after the strike's chaotic climax, Harry Orchard, a former member of the Altman WFM local, confessed that he had murdered Frank Steunenberg in Caldwell, Idaho, December 30, 1905. A sheep rancher in 1905, Steunenberg was the governor of Idaho during the Coeur d'Alenes strike of 1899. Orchard implicated an "inner circle" of the WFM in the Steunenberg murder, particularly WFM President Charles Moyer, Secretary-Treasurer W. D. (Big Bill) Haywood, and George Pettibone, a Denver merchant who had been a WFM activist in the Coeur d'Alenes. The three were abducted to Idaho, imprisoned for over a year and a half, and finally acquitted.[25]

Orchard also alleged that the "inner circle" hired him to set the explosives at the Vindicator Mine and the Independence Depot. Whether Orchard in fact blew up the depot, whether he acted alone, and who hired him remain contested questions.

The answers to these questions have proved difficult for the best-intentioned historians because sources and memories were among the spoils of the strike. The daily newspaper owned by the local labor movement folded just as the strike began, after which the *Victor Record* was the only remaining pro-union paper in the district.

One night the militia rounded up the *Record*'s workforce, and took them to the military bullpen. Emma Langdon, a union printer, sneaked in, printed the paper, and delivered it to the militia, the ink still wet, its headline blazing, "Slightly Disfigured But Still In the Ring."[26] The *Record* offices were sacked in the strike aftermath; it then abandoned its pro-union stance.

The *Victor Record* force in the bullpen, (left to right) F. W. Langdon, Linotype Operator; Charles G. Langdon, Linotype Operator; George E. Kyner, Managing Editor; W. L. Sweet, Circulator; H. J. Richmond, Foreman; Soldiers with loaded Winchesters on guard. *From* The Cripple Creek Strike, *by Emma Langdon, Special Collections, Pikes Peak Library District.*

Other records were seized or destroyed when mobs poured coal oil on them as they attacked the union halls in the June 6 rioting.[27] Some memories left with the leaders forced from the district; others died before their stories were recorded. Controlling sources and memories became part of the history of the strikes and their interpretations.

Competing histories appeared by mid-strike. The Colorado Mine Owners' Association published *The Criminal Record of the Western Federation of Miners from Coeur d'Alenes to Cripple Creek* in 1904, charging the WFM with virtually every crime committed in the gold camps for a decade. The WFM countered with its *Category of Crimes of the Operators' Association: 851 Murdered in Less Than Four Years,* which blamed all mining-related deaths on owners' negligence. Emma Langdon published her pro-labor history in 1904 and 1905. The counter-narrative appeared in the first academic history, Benjamin McKie Rastall's *The Labor History of the Cripple Creek District,* published in 1908 from his 1906 University of Wisconsin dissertation.[28] A Colorado Springs resident, Rastall had close ties to local business; he somehow secured some of the few surviving union records, which remain at the Wisconsin Historical Society.

For Rastall, the strike was unreasonable because the miners struck innocent mine owners with whom Rastall maintained they had no grievance; he ignored the vertical integration of mines and smelters. He characterized the district as a "Little Mining Monarchy, shut off from the rest of the world by a high mountain range . . . with its own separate political organization, the power in the hands of the miners, the unions controlled by a minority, the strike power in the hands of a clique." It was an "anomaly," "a very recent frontier district, of a highly developed industrial center, which has kept most of its frontier aspects."[29] The mine owners' forcible takeover of the local government and the courts thus civilized the frontier.

> No discredit is to be attached to the mine owners for opposing a solid front to the attack of the Western Federation of Miners, nor for straining every nerve to break the strike. To have submitted would have been to admit a condition of subserviency impossible to men of strength and courage.[30]

Rastall insisted that everyone he knew in the district knew that the union had blown up the depot. He and I certainly talked to different people, but his version dominated local histories that focused on the lovable antics of fun-loving, hard-drinking young mine owners, especially those who won control of the district after the strike.[31]

For almost a century historians have debated these strikes and what made the miners who waged them so weird. Seeking to explain the union's socialist endorsements and the violence of many western miners' strikes, historians concluded variously that the miners' politics were the product of the frontier;[32] of brutal strike experiences;[33] of the lack of a local middle class;[34] of rapid industrialization;[35] of the dangers of underground work;[36] of loss of control to management;[37] of diminished opportunities for self-employment;[38] of failure to secure an eight-hour law in Colorado;[39] or of the defeats of the Populist Party and the Knights of Labor.[40] For some, the strikes became insignificant because they were not ordinary daily occurrences. As one historian put it, their significance rested in the false assumption that "the WFM's defeat in Cripple Creek was merely the experience of Tincup, Silver City, Lump Gulch, and Tonopah writ large."[41]

Perhaps not. One could argue—and I have—that Cripple Creek was, both materially and symbolically, more significant than Tincup in shaping working-class agendas and in the lived experience of tens of thousands of miners and their families; that violent confrontations did not have to occur in every mining town to influence workers' understandings; that when mine owners broke strong unions, the outcomes shaped relationships throughout the industry.[42]

But rather than debate the debate further, let's return to Cripple Creek "writ small"—not Tincup or Lump Gulch, but John Welch and his descendants who preserved his memory. Their oral histories linked an extended family of Welches, McConaghys, and Dorans. Hannah Welch's parents, Edward and Catharine Doran, were born in Ireland in the early 1830s. They married young, probably during the disastrous famine years of 1845–1849, left Ireland for Wales where their first son was born in 1852, and then for Shullsburg, in Wisconsin's lead mining region. Edward worked as a lead miner. Catharine bore twelve children in nineteen years, nine of whom survived. All of the Doran children attended school well into adolescence, an important achievement for Catharine, who could neither read nor write. By the 1880s, lead mining declined, and the new silver mines around Leadville drew many Shullsburg miners, including the Dorans.[43]

Most of the Doran sons mined; the other daughters, too, married miners. Kate Doran married James McConaghy, an Irish immigrant hoist engineer. Their daughter May was born in Leadville November 25, 1890. Three months later Kate died, and in 1893 James took three-year-old May to Victor, in the Cripple Creek District. James' brother John McConaghy moved there, too. So did his brother-in-law Ed Doran, with his wife Mary.[44]

In 1893, John and Hannah Doran Welch followed. After the first strike, they moved to a new home they built in Goldfield. Kathleen was born there in 1895, then Edward, Annie, and young John. John and Hannah Welch told the next generation of Welches, Dorans, and McConaghys their stories of the strikes and of the labor community. Kathleen Welch Chapman fondly recalled union benefits as part of what her father meant when he said that unions were "for people."

> Now, if there was a family that was very hard up or very tight for money, why [the unions would] help them out. . . . I can remember people in Victor coming to Goldfield when Papa'd maybe have a cold or be sick or couldn't go to work or something. There'd be maybe a half a dozen men come at night to see if he was getting along and if he needed anything.[45]

One of the first tasks of local unions was to provide for sickness and death. Unions paid sick benefits, cared for widows and orphans, organized funerals, shipped bodies out of the district for burial and hired doctors and nurses for their members.[46]

The unions organized social life as well, hosting balls, socials, picnics, and holidays. The two big celebrations were the Fourth of July and Labor Day. To little Kathleen Welch, the annual Labor Day picnic at Pinnacle Park was a wonderland of merry-go-rounds, ball games, horseshoes, picnics shared with other

Union parade led by the Victor Union No. 32 of the Western Federation of Miners Association, ca. 1890s. *Photograph from Special Collections, Pikes Peak Library District.*

union families, speeches by the governor of Colorado and other "people like that you know, that was really worthwhile."[47]

Labor Day began with a parade. An estimated three thousand union members marched in the 1899 parade, five thousand in 1901 and 1903. John Welch marched with his religious fraternal association, the Knights of St. John, not his union local. He "always wore his uniform," Kathleen remembered, "and, oh, them were great."[48] Kathleen Chapman did not remember whether or not her mother belonged to the WFM Women's Auxiliary, but Hannah was, she said, "always quite a church member," and she remembered with great fondness the local priest who sided with the union during the strike.[49]

After 1894 John had generally cordial relations with his employers, the Woods Investment Company. The Woods family were Baptists who believed, apparently, in the stewardship of wealth. Unlike most district owners, they invested some of their wealth in the district and were widely perceived as generous and egalitarian for their contributions to working-class social life. The family built Pinnacle Park at Cameron, a vast picnic ground, zoo, and amusement park, which the Woods regularly donated for

Labor Day. Soon after the park opened, the Woods family gave a "grand picnic" for the district, and they hosted an annual New Year's reception and entertainment for the employees of all their mines.[50]

The Woods Investment Co. built a clubhouse for its employees, opposite their Gold Coin Mine in Victor. The Club was dedicated in February 1900 at an elaborate party with dancing and a banquet. John and Hannah Welch were there. Their daughter remembered the night vividly.

Gold Coin Club, 1980. *Photograph by Myron Wood, from Special Collections, Pikes Peak Library District.*

> Mama and Papa went. I can remember how nice they looked. They got a baby-setter to come in and set with us. . . . We lived in Goldfield, and they went up to Victor to the opening. And . . . they said they had such a good time. Well, they danced and they had a big banquet.[51]

Not surprisingly, Kathleen Chapman remembered parades, baby-sitters, and union committees that came to her house when her father was ill, but not what happened at the union hall or John Welch's political affiliations. I don't know whether he shared the Populist politics of many Cripple Creek miners in the 1890s, whether he, like an influential minority of the union leadership, joined the Socialist Party after it was formed in 1901.

Whatever John Welch's politics, he shared a great deal with his co-workers and with the union leadership. Many were, like him, married homeowners in their thirties and forties, who lived in the smaller towns like Goldfield.[52] Like him, they had amicable relations with employers as long as they felt employers treated workers with respect. Otherwise, they struck, or boycotted firms that were unfair to workers in other crafts, those who sold scab goods, or who refused to honor labor's demands for wages, hours and working conditions.[53]

There were more than two Cripple Creek strikes. Between the major conflicts in 1894 and 1903–1904, a number of local disputes were settled fairly easily. During 1899–1903, the *Press* reported at least thirty-four minor strikes, primarily over wages and working conditions, most of which were quickly negotiated.[54]

Strikes were a last resort. Generally, unions simply announced their wages and hours and boycotted businesses that did not meet their demands. During 1901 alone, the *Press* advertised boycotts of more than twenty products or firms. Using the *Press* to publicize banned goods and establishments, organized labor established six o'clock closing for retail clerks, prevented laundries from shipping clothes out of the district to non-union shops, made boarding houses and restaurants hire union help, established the six-day week for cooks and waiters, eliminated non-union products from saloons, groceries, and other retail houses—in short, enforced demands for union recognition, wages, and hours for most district workers.[55]

Boycotts worked best for trades that provided local services, like waitresses, cooks and bartenders, or for workers who produced consumer items, like cigars, beer, or clothing. They

Day Shift on the Calaconda, October 7, 1897, Victor, Colorado. A handwritten note on back reads "1. John Murrah. Pump man on Golden Cycle. 2. Joe Acton [name crossed out in pencil] on Annie J." *Photograph from Special Collections, Pikes Peak Library District.*

were less effective for miners, who could not have gotten far with a working class gold boycott. Instead, WFM organizers often depended on social pressure, like checking miners' union cards as they went on shift. John Welch was probably there when WFM officials checked cards at the Gold Coin in May 1900. Only one miner resisted, saying that he "cared nothing for the miners' union or its representative." The rest of the miners refused to ride down in the cage with him, creating a stalemate that was resolved only when the superintendent paid the man's dues.[56]

We don't know for certain that John Welch refused to ride down in the cage with a non-union man, though he worked at the Gold Coin, and from his daughter's account we would expect him to be with the union majority. Nor do I know how he voted. I do know that he walked out with other strikers when the Woods were forced to fire their union workers, and stayed out for the duration of the strike. Despite common pictures of married homeowners as working-class conservatives, and of radicals as young, single, rootless hotheads, in fact the single men could simply leave during strikes. The stalwarts during the 1903–1904 strike were, like John Welch, married men and homeowners, who had the most to risk and the most to lose.[57]

However romantic strikes may seem in history books, they are simply hard in practice. For fifteen months John Welch stuck it out, probably like many others working only on union mines, making do with strike benefits. He was considered sufficiently important to be slated for deportation in the brutal aftermath of the strike, and left the district, with his brother-in-law Ed Doran, one step ahead of the militia.

Kathleen Chapman remembered the second strike; the militia camped on the hill above her home, standing with other children to wave at the trainloads of deportees. She recalled that her parents helped a pro-union soldier desert his post and sneak out of the district. And she remembered that:

> Papa had to get ready, him and Uncle Ed, and left, things got so bad. Somebody come and give 'em the word that they were going to get them the next trainload they took out. . . . And so they got ready, and they left in the middle of the night, the two of them. And they went to Leadville, and went to work.[58]

After John left, Hannah Welch maintained their union allegiance. Kathleen's cousin May Wing fondly recalled the strike heroism of her grandmother Catharine Doran and her aunts, Mary Doran and Hannah Welch.

Tents scattered throughout small mine buildings at Goldfield, Colorado. Identified as "Colorado National Guard Camp near Cripple Creek during the miners strike." *Photograph courtesy of Elizabeth Ogden, from Special Collections, Pikes Peak Library District.*

> I did have an uncle that lived in Goldfield. His name was Ed Doran. And Grandma was there, Aunt Mary was expecting. So Uncle Ed, of course, got out. Well, he got word that they were going to raid the house. Of course, the militia had the way of doing, in the middle of the night they'd come after the men and take 'em to the bullpen and then they'd send 'em out. But Uncle Ed got word that they were going to come that night, so he left. Grandma never said just exactly where he went or where he hid. But she was there when they came, and they tore that house to pieces! They even looked in the breadbox. Well, that got Grandma's goat. And she said, "Shure'n he's not little enough to put in the breadbox! Now every God blesset one of you get out of this house and leave this woman alone!" She said, "You can see what condition she's in. Now," she said, "git!"

> Then I had another aunt that lived there. Her name was Hannah Welch. . . . And she had two great big butcher knives, and she kept those knives razor sharp. And she always said if one of those militia men ever come in her house in the middle of the night, they'd leave with less than they brought in![59]

Finally, though, Hannah Welch took the children to join John in Leadville, after the militia harassed her son Tom and her nephew, Eddie Bulger, as the boys were going fishing. Kathleen recalled:

Snowfall at the National Guard Goldfield camp with mine buildings in the distance. Identified on the back as "1903? . . .Colorado National Guard, Cripple Creek, Colorado." *Photograph courtesy of Elizabeth Ogden, from Special Collections, Pikes Peak Library District.*

> [T]he militia drove up there and stopped them. And they wanted to know where they were going. They . . . said, "Oh, no, you're not. You're hauling out food for the union men someplace." "Oh, no," Eddie said, "we're not. We don't have anything, only our lunch." Anyway, they pulled the two kids down off of the horse. Eddie was about six years older than Tom. And pulled them down, took their lunch, throwed it down on the grass.

> Took the saddle off the horse, and every little nook and corner of that saddle they went through to see if they were taking something. . . . [A]nd you know that made a wreck out of Tom. He was about, I'd say, ten, eleven years old, something like that. And he was just a wreck. He couldn't sleep at night. . . . He'd scream in his sleep, you know. So Mama said, "We'll I can't stand that." So she packed us up and took us to Leadville, where Papa was, until the darned thing would quiet down or get over. And we were up there for over a year and a half. Then we came back to the district.[60]

Hannah and the children stayed in their home in Goldfield. John Welch worked in various mining communities, and visited his family when he could, because, Kathleen said, "a union man couldn't get a job and he wouldn't work on them scab cards for anything." After a mine cave-in in Minturn crushed his chest and injured his legs, John Welch came home. Unable to mine underground, he worked as a watchman at the Trilby, a small mine that resisted the Mine Owners Association.

Finally, though, Welch became too ill to climb the hill to the mine or to do much work. His son Tom left school after eighth grade to take his father's place. Just as John Welch entered the workforce when *his* father died in a Vermont coal mine, Tom Welch, born in a tent during the Bull Hill strike, became the family breadwinner. In time, Kathleen contributed her wages

Deported men being driven across the New Mexico State Line. Antonito Mountain in the distance. *From* The Labor History of the Cripple Creek District, *by Benjamin McKie Rastall, Special Collections, Pikes Peak Library District.*

as a housekeeper and laundry worker. John Welch died of his injuries at age forty-eight. His children remembered the union benefits, the social support, the vibrancy and the power of the union community their father had built and cherished. They knew what they had lost.[61]

The casualties of the labor wars went far beyond individual lives. The Cripple Creek working class also lost the collective records of their hopes and history. To work in Cripple Creek, miners had to hide any hint of union allegiance. By the 1970s, the people who remembered the union era were mostly elderly women whose fathers had been union miners, who married men too young to have begun mining before 1904 or to have been union members. But they could still be blacklisted, so it was dangerous to talk about organized labor and the strike. The mines closed in the early 1960s, and elderly residents felt free to tell their stories—even to a prying young historian.

The history they wished to preserve was not a nostalgic past, or a tale of tragic loss. It was about complicated people who lived complex, mundane lives. They sought neither romance nor riches, but a community that they could shape, a place where work would be honored and workers dignified.

The miners of the Cripple Creek District risked a great deal daily, underground in the mines, and in the strike we met to commemorate. They risked for what drives most of us—to keep their children in school, to work in dignity, to go home to a house that they owned. Years later, the survivors recalled the strike defeat in terms of domestic loss. Asked what difference the second strike made for working people, Kathleen Chapman replied:

> Oh my, it changed it awful! . . . [Y]ou should have seen the homes that was up there in Goldfield. They were beautiful, some of them. They were just gorgeous. Well, people just got up and left, you know, and left their homes like it was, and lots of them never come back to them. People would go in and destroy the homes, and pick up what they wanted, and the houses went to wrack, and they tore them down, and—oh, it just, it just used to make me sick all over.[62]

The deserted homes of the Cripple Creek District represented both actual wrenching losses and, metaphorically, the heart of a labor community that sustained them. Like the mute decaying houses, the mundane achievements of a labor community were both messier and more profound than a debate about violence or accommodation, socialism or mainstream politics. The nitty-gritty daily accommodations of the union era, the power to conduct heated political debates that remained unresolved, the empowerment of acting in such local arenas—these may well have been, for John Welch, and for countless others in Cripple Creek, Lump Gulch, and maybe Tincup, too—the measure of what they achieved and what they lost.

Union Hall, Victor Miners Union No. 32, after its occupation by the militia. *From* The Labor History of the Cripple Creek District, *by Benjamin McKie Rastall, Special Collections, Pikes Peak Library District.*

The meaning of the strike centennial is inextricably bound to history and memory. The local versions of the 1903–1904 strike depended on peoples' allegiances in what became a company town. The versions of events that people have accepted, and still passionately argue, have been shaped by published histories, by personal experience, and by political and ethical allegiances. Depending on the cast and the plot, the centennial marked the

Men stand on the tracks surveying the damaged Independence Depot building after a bomb exploded, killing 13 non-union miners. Dated June 6, 1904. *Photograph by Schedin & Lehman, from Special Collections, Pikes Peak Library District.*

triumph of beleaguered manly mine owners, the sacrifice of human rights to property rights, the losses of a labor community.

I have published a version of these events, including circumstantial evidence that might support labor's claims that owners were responsible for the carnage at the Independence Depot.[63] But I don't think anyone yet really knows what happened. Perhaps when I retire I can follow all the dangling leads, but the evidence may no longer exist to resolve that issue.

Rather, let me suggest briefly some of what the strike was and was *not* about. It was certainly about basic constitutional rights: freedom of speech, freedom of assembly, the right to *habeas corpus*, the right to organize, the right to control local communities and daily discourse, the right to elect public officials. It was not a simple contest between capital and labor, but also among competing employers who disagreed about recognizing organized labor and negotiating with their workers. It was not an isolated event. The Colorado Labor Wars were part of a connected pattern of labor relations throughout the mining West, from the Coeur d'Alenes in 1892 through Ludlow in 1913–1914, and the Bisbee deportations in 1917, where violent incidents justified state-sponsored violence to crush organized labor. It was not about socialism, though some union members were socialists, but about the survival of a labor movement and a working-class culture that supported debate and dissent.

And whatever else this history is about, it is not about who was most violent.

Whether or not they staged the fatal accidents at the Vindicator, Stratton's Independence, and the Independence Depot, the Mine Owners Association, Citizens' Alliance, and the militia fomented the Victor riot that caused two deaths, destroyed union property, arrested and deported people charged with no crimes, deposed elected officials, and at least tacitly approved assaults and whitecappings. They deported over two hundred citizens hundreds of miles from their homes under military guard and left them in the middle of nowhere at the state line without food or shelter. Even if one believes union members staged the bombings, it is still difficult to determine blame as the local unions and even the Women's Auxiliary were heavily infiltrated by mine owners' detectives, who in other circumstances instigated violence. Even Harry Orchard, whose truthfulness should be subject to doubt, said that he worked for the Pinkerton Detective Agency during the Cripple Creek strike.[64]

I do not admire violence or condone it. The strike violence magnified a prevalent culture of masculine violence, in which prizefights were popular entertainment, and miners lived with the constant threat of violence and death underground. Both sides probably used violence. The MOA paid to blowup assay offices in 1902 to try to stop high-grading (selling stolen high-grade ore).[65] Union men beat up scabs and ran "non-white" workers out of camp. The strike record testifies that men on all sides habitually prepared for violence, and either carried guns or hired them. Even Hannah Welch "kept her knives razor sharp."[66]

In terms of the historical outcome, the question of who actually planted the dynamite at the Independence Depot is less important than whose explanation was accepted, and how that justified the consequences. Perhaps the most eloquent testimony to the importance of the Colorado Labor Wars lies in recent attempts to silence their history.

The desecration by vandals of the Ludlow Memorial literally dismembered labor's commemoration of a later Labor War that involved some of the same actors, the same militia members, the same hired gunmen. Efforts to suppress histories of resistance and dissent have obvious resonance today as we again debate the right to hold unpopular political opinions, and consider which historical sites should be preserved and which torn down to erase unpleasant memories.[67]

The strike commemoration, like the strikes themselves, is about who controls history. The story is always in danger of being lost, buried, or trivialized. May Wing's challenge remains urgent, in a world where the daily news preserves some memories and erases others. We will continue to debate the strike events and their consequences. But the most urgent testimony to the importance of this centennial comes from John and Hannah Welch, Kathleen Chapman, May Wing, and others who lived the history that they could tell and told the history they risked to make.

◊◊◊◊

Notes

I want to thank the staff of the Pikes Peak Library District for organizing the Cripple Creek Strike Commemoration, and especially Chris Nicholl, without whom it wouldn't have happened. I also thank Ed and Cherrie Hunter for help arranging events in the Cripple Creek District. And, as always, I thank the members of the extended Wing, Pryor, and Chapman families for sharing their memories and more with me. Portions of this article appear in my book, *All That Glitters: Class, Conflict, and Community in Cripple Creek*, and appear here with the permission of the University of Illinois Press; I am grateful.

1. *Webster's New World Dictionary of the American Language, College Edition* (Cleveland & New York, 1962), 918, 1230, 293, 420, 689.

2. See Benjamin McKie Rastall, *The Labor History of the Cripple Creek District: A Study in Industrial Evolution*, Bulletin of the University of Wisconsin No. 198, Economic and Political Science Series vol. 3, no. 1 (reprint, Madison, Wisconsin: 1908); Emma F. Langdon, *The Cripple Creek Strike: A History of Industrial Wars in Colorado* (1904–05; reprint, New York: 1969); William D. Haywood, *Bill Haywood's Book: The Autobiography of William D. Haywood* (New York: 1929; reprint, 1974), 108–73; Vernon H. Jenson, *Heritage of Conflict: Labor Relations in the Nonferrous Metals Industry Up to 1930* (Ithaca, N.Y.: 1950), 118–59; Marshall Sprague, *Money Mountain: The Story of Cripple Creek Gold* (New York: 1953); Melvyn Dubofsky, *We Shall Be All: A History of the Industrial Workers of the World* (New York: 1969), 49–55; George G. Suggs, Jr., *Colorado's War on Militant Unionism: James H. Peabody and the Western Federation of Miners* (Detroit: 1972), 84–117; Elizabeth Jameson, *All That Glitters: Class, Conflict, and Community in Cripple Creek* (Urbana: 1998), 199–252.

3. See for examples Sprague, *Money Mountain*; Marshall Sprague, *Newport in the Rockies: The Life and Good Times of Colorado Springs* (Denver: 1961); Frank Waters, *Midas of the Rockies: The Story of Stratton and Cripple Creek* (Chicago: 1937); Mabel Barbee Lee, *Cripple Creek Days* (Garden City, N.Y.: 1958), 186–224; Thomas J. Noel and Cathleen M. Norman, *A Pikes Peak Partnership: The Penroses and the Tutts* (Boulder: 2000); Leland Feitz, *Myers Avenue: A Quick History of Cripple Creek's Red-Light District* (Denver: 1967); Caroline Bancroft, *Six Racy Madams of Colorado* (Boulder: 1965).

4. See Selig Perlman and Philip Taft, *History of Labor in the United States, 1896–1932*, vol. IV, 173-78, 191-203; John R. Commons, et al., *History of Labor in the United States* (New York: 1935), 174–78, 194–203; Jenson, *Heritage of Conflict*, 38–53, 118–59; Suggs, *Colorado's War*; Dubofsky, *We Shall Be All*, 36–56.

5. Interview with Leslie and Ruby Wilkinson, Cripple Creek, Colorado, September 7, 1975; interview with Beulah Pryor, Colorado Springs, Colorado, May 6, 1979; interview with Beulah Pryor and May Wing, Colorado Springs, Colorado, February 17, 1979; interviews with S. C. Gatheridge, Denver, Colorado, February 21, 1979, and March 7, 1979; interview with Kathleen Chapman, Wheat Ridge, Colorado, April 27, 1979.

6. Interviews with May Wing, Boulder, Colorado, March 6, 1976; Victor, Colorado, October 21, 1978; Colorado Springs, Colorado, February 16, 1979.

7. May Wing interview, February 16, 1979.

8. See Jameson, *All That Glitters*, 62–86, 258–69.

9. See Perlman and Taft, *History of Labor in the United States, 1896–1932*, 174–78, 194–203; Jenson, *Heritage of Conflict*, 38–53, 118–59; Suggs, *Colorado's War*; Dubofsky, *We Shall Be All*, 36–56; Langdon, *The Cripple Creek Strike;* Jameson, *All That Glitters*, 199–252.

10. *Webster's*, 689, 918, 420, 1230, 293.

11. The monument was dismembered May 7, 2003. See "Vandals Attack Ludlow Memorial to Slain Miners," *Pueblo Chieftain*, May 10, 2003.

12. For a more complete quantitative portrait of the Cripple Creek working class see Jameson, *All That Glitters*, 21–40.

13. Kathleen Chapman interview.

14. Ibid. In 1894, the Cripple Creek District was part of El Paso County, the county seat of which was Colorado Springs. In 1899, the Colorado legislature carved Teller County out of parts of El Paso and Fremont Counties, and Cripple Creek became the county seat. The 1903–1904 strike pitted the power of the miners' local governments in Teller County against the combined power of the mine owners' home base in El Paso County, and that of the state.

15. For competition among the owners, see Jameson, *All That Glitters*, 41–61, 199–235.

16. *Victor and Cripple Creek Daily Press*, February 15, 1903; February 18, 1903; February 19, 1903; February 20, 1903; February 21, 1903; February 22, 1903; February 26, 1903; March 1, 1903; March 4, 1903; March 6, 1903; March 7, 1903; March 11, 1903 (hereafter cited as *Press*); U.S., Congress, Senate, Commissioner of Labor, *A Report on Labor Disturbances in the State of Colorado from 1880 to 1904, Inclusive, with Correspondence Relating Thereto*, S. Doc. 122, 58th Cong., 3d Sess. (Washington, D.C.: 1905), 112–16, (hereafter cited as *Labor Disturbances*); *Official Proceedings of the Eleventh Annual Convention, Western Federation of Miners* (Denver: 1903), 115–19; Suggs, *Colorado's War*, 44–117, esp. 76–77, 81–83.

17. Suggs, *Colorado's War*, esp. 65–83.

18. *Labor Disturbances,* 165–66, 170–75; *Denver Post,* December 9, 1903. Peabody appointed Bell after the Colorado City strike began, on February 20, 1903; *Labor Disturbances,* 147–48; Suggs, *Colorado's War,* 81.

19. *Labor Disturbances,* 189-93, 220–23; Defense Attorney Richardson's notes, affidavits, and related material 1906–1909, State of Idaho vs. Steve Adams, Western Federation of Miners/International Union of Mine, Mill and Smelter Workers Collection, Western Historical Collections, University of Colorado, Boulder (hereafter cited as Richardson Notes), 75; "The Situation in Colorado," *Miners' Magazine,* November 26, 1903, 7; *Miners' Magazine,* February 4, 1904, 3–4; "Western Federation Notes," *Miners' Magazine,* February 4, 1904, 9.

20. *Labor Disturbances,* 181–87; *Miners' Magazine,* September 17, 1903, 9; *Pueblo Labor Advocate,* September 18, 1903; "The Governor of Colorado," *Miners' Magazine,* September 24, 1903, 4–5; "The Cripple Creek Situation," *Miners' Magazine,* October 1, 1903, 7; Dubofsky, *We Shall Be All,* 50.

21. See for instance postings on the H-Net Labor History Discussion List, October 14, October 25, October 27, October 30, November 3, 2002. See also postings September 11, 1998, and on H-West, October 8, 1996.

22. See Suggs, *Colorado's War,* 110–14; Jameson, *All That Glitters,* 218–25; *Labor Disturbances,* 260–68, 274–78, 285–86, 295, 306–08, 313–14.

23. *Labor Disturbances,* 285–86; "The Situation in Colorado," *Miners' Magazine,* June 16, 1904, 10; "The Situation in Colorado," *Miners' Magazine,* June 23, 1904, 9; *Miners' Magazine,* July 21, 1904, 4; "The Situation in Colorado," *Miners' Magazine,* July 28, 1904, 9–10; "The Situation in Colorado," *Miners' Magazine,* February 16, 1905, 11.

24. "The Situation in Colorado," *Miners' Magazine,* November 17, 1904, 10–11; *Denver Clarion Advocate,* November 11, 1904, 1.

25. See Dubofsky, *We Shall Be All,* 97–105; Harry Orchard, *The Confessions of Harry Orchard* (New York: 1907), esp. 68–109, 129–48.

26. "The Cripple Creek Situation," *Miners' Magazine,* October 8, 1903, 6.

27. For the destruction immediately following the Independence Depot explosion, see *Labor Disturbances,* 249–51; Suggs, *Colorado's War,* 111; Operative J. N. L. Reports, Victor, June 6, 1904, Pinkerton Detective Agency Reports, University of Colorado, Western Historical Collections, Boulder; Edward Verdeckberg Papers, Colorado State Historical Society, Denver, 97.

28. Colorado Mine Owners' Association, *The Criminal Record of the Western Federation of Miners from Couer d'Alenes to Cripple Creek* (Colorado Springs: 1904); *Category of Crimes of the Operators' Association: 851 Murdered in Less Than Four Years* (Denver: 1904); Langdon, *The Cripple Creek Strike;* Rastall, *Labor History.*

29. Rastall, *Labor History,* 72–63.

30. Ibid., 152.

31. Sprague, *Money Mountain*.

32. Jenson, *Heritage of Conflict*.

33. Perlman and Taft, *History of Labor in the United States, 1896–1932*; Melvyn Dubofsky, "The Origins of Western Working Class Radicalism, 1890–1905," *Labor History* 7 (Spring 1966): 131–54; Dubofsky, *We Shall Be All*.

34. Louis Filler, *The Muckrakers: Crusaders for American Liberalism* (New York: 1939; reprint, Yellow Springs, Ohio: 1950), 217–18; Clark Kerr and Abraham Siegel, "The Interindustry Propensity to Strike," in *Industrial Conflict*, eds. Arthur Kornhauser, Robert Dubin, and Arthur M. Ross (New York: 1954), 189–212.

35. Dubofsky, "Origins of Western Working-Class Radicalism" and *We Shall Be All*; Richard E. Lingenfelter, *The Hardrock Miners: A History of the Mining Labor Movement in the American West, 1863–1893* (Berkeley: 1974), 227–28.

36. Kerr and Siegel, "Interindustry Propensity to Strike"; Ronald C. Brown, *Hard-Rock Miners: The Intermountain West, 1860–1920* (College Station, Texas: 1979).

37. Brown, *Hard-Rock Miners*.

38. John H. M. Laslett, *Labor and the Left* (New York: 1970).

39. Mark Wyman, *Hard Rock Epic: Western Miners and the Industrial Revolution, 1860–1910* (Berkeley: 1979).

40. William Preston, "Shall This Be All?: U.S. Historians Versus William D. Haywood *Et Al*." *Labor History* 12 (Summer 1971): 435–53.

41. James C. Foster, "Quantification and the Western Federation," *Historical Methods Newsletter* 10 (Fall 1977): 141–48; quote from Wyman, *Hard Rock Epic*, 244. See also John Ervin Brinley, Jr., "The Western Federation of Miners" (Ph.D. dissertation, University of Utah, 1972); Richard H. Peterson, "Conflict and Consensus: Labor Relations in Western Mining," *Journal of the West* 12 (January 1973): 1–17.

42. Jameson, *All That Glitters*. I most explicitly addressed Wyman's statement in a paper I presented at the Western History Association Conference in October 1992, "The Miners of Cripple Creek: Accommodation and Resistance, 1894–1904."

43. The interconnected Doran, McConaghy, and Welch family histories were provided by John and Hannah Welch's daughter, Kathleen Chapman, and James and Catharine McConaghy's daughter, May Wing. Kathleen Chapman interview; May Wing, interviews. Further information from Manuscript Census, town of Shullsburg, LaFayette County, Wisconsin, 1860; Manuscript Census, town of Shullsburg, LaFayette County, Wisconsin, 1870; Manuscript Census, town of Shullsburg, LaFayette County, Wisconsin, 1880;

Manuscript Census, town of Leadville, Lake County, Colorado, 1900; *The* (Shullsburg) *Pick and Gad,* June 19, 1884.

44. May Wing interview, March 6, 1976.

45. Kathleen Chapman interview.

46. In 1896, for instance, the Anna Lee shaft of the Portland Mine caved in, killing eight miners. Victor Miners No. 32 buried victim Michael McGuirk, and shipped the body of Pat Mee to Denver for burial, and the body of Thomas H. Harman to Madison, Wisconsin. *Press,* April 6, 1900; *Colorado Springs Gazette,* January 13, 1896, January 30, 1896; *Ledger,* 1894–1903, Victor Miners' Union No. 32, U.S. Mss 14A, Wisconsin State Historical Society, Madison, Wisconsin; *Press,* October 5, 1902, August 1, 1901.

47. Kathleen Chapman interview.

48. *Press,* September 5, 1899; September 3, 1901; Haywood, *Bill Haywood's Book,* 133; Kathleen Chapman interview.

49. Kathleen Chapman interview.

50. Ibid.; *Press,* August 9, 1899; January 3, 1900; January 9, 1900.

51. *Press,* June 27, 1899; February 23, 1900; "The Gold Coin Club House," *Miners' Magazine,* July 1900, 13–14; Kathleen Chapman interview.

52. Jameson, *All That Glitters,* 78–84.

53. Ibid., 68–72.

54. *Press,* passim; Jameson, *All That Glitters,* 70.

55. Jameson, *All That Glitters,* 70–71.

56. Ibid., 73. See *Press,* May 9, 1900; May 15, 1900. Such tactics were common. The next day, Kelly and E. J. Campbell, secretary of Cripple Creek Miners No. 40, visited the Independence Mine. See *Press,* May 16, 1900, July 22, 1900.

57. Jameson, *All That Glitters,* 235–39.

58. Kathleen Chapman interview.

59. May Wing interview, October 21, 1978. Kathleen Welch Chapman also knew these stories. Kathleen Chapman interview.

60. Kathleen Chapman interview.

61. Ibid.

62. Ibid.

63. See Jameson, *All That Glitters,* 227–33.

64. *Labor Disturbances,* 189–91; Richardson Notes 27, 75; Dubofsky, *We Shall Be All,* 98; George E. Dickson, letter to Clarence Darrow, April 25, 1906, in Richardson Notes, 53. Dickson was a Chicago attorney who did investigation for the defense during the Adams and Haywood trials. He got the story from G. L. Brokaw, who was jailed in Spokane for getting money under false pretenses, and who said that he was with Orchard for ten days before May 26, 1905. See also George E. Dickson to Edmund F. Richardson, May 6, 1906, Richardson Notes, 55.

65. *Press*, February 25, 1902, March 11, 1902; *Colorado Chronicle*, February 26, 1902.

66. See Chapters 6 and 8, Jameson, *All That Glitters;* May Wing interview, October 21, 1978.

67. When I wrote these words, I was thinking about the decision to tear down the Abu Ghraib prison. There is a larger issue of what defines a historic site, and which memories should be claimed collectively.

Miners posing in a rough tunnel. Identified as "In Breast of Tunnel 4500 Feet from Portal." *Photograph by Schedin & Lehman, from Special Collections, Pikes Peak Library District.*

Striking Western Federation of Miners, some of whom were deported from the Cripple Creek District in 1904. Clockwise from top left: Delfido Gonzales (from Springer, New Mexico, deported), Jerry Kelly (from Victor, Colorado, Secretary-Treasurer of the Victor Miners' Union No. 32, deported), Harry Robinson (from Altman, Colorado, deported), L. E. Krotz (from Victor, Colorado, deported) and J. D. Stahl (center, origin unknown). *Photographs from Special Collections, Pikes Peak Library District.*

Pat Mullaney appears on the list of striking miners. A note on the card reads, "Beat up in Bull Pen." *Photograph from Special Collections, Pikes Peak Library District.*

"Meet Me at the Ballot Box": Women Voters, Striking Miners, and Colorado Labor Politics, 1893–1904

Marcia Tremmel Goldstein

"The ballot is the only remedy
for those who toil and delve.
Meet me at the ballot box
. . . and cast your ballot
for your own. . . . It is the only
salvation of the laboring people."

Emma F. Langdon,
The Cripple Creek Strike
(1904–1905)[1]

This quotation from Cripple Creek journalist, union activist, and historian Emma Langdon, captures a key message we can take away one hundred years after the dramatic events of the Victor and Cripple Creek labor war of 1903–1904. Langdon's call to "meet me at the ballot box" came in the wake of disastrous violence and extralegal attacks on striking miners in the District at the hands of Republican governor James Peabody. At the time, working people in Colorado had suffered a deep loss of faith in the electoral system. However, Langdon's words reflect the political strategy that was adopted by the labor movement in the wake of Peabody's anti-union attacks: unite all fair-minded Colorado citizens to defeat the politicians who controlled the citizens by fear and force. Her message reveals that even the most radical, socialistic labor leaders and voters aspired to "work within the system"—to test the principles which our democracy promises, but does not always deliver.

This presentation is part of a work in progress, in which I have been researching the particular role of newly enfranchised women voters and political activists within our state's volatile, polarized political climate a century ago. I will argue that, prompted by a sense of fairness and the need for a more equitable community, a large segment of women voters and activists from all walks of life supported labor's cause at the polls and in the legislature in the 1890s and early 1900s. Labor leaders reciprocated by getting out the vote for important reform measures for working women such as the eight-hour day for women and the "Mother's Pension" (an early form of welfare). Women's early enfranchisement and the strength of the Rocky Mountain labor movement allowed Colorado to become a political "hot-house," where the power of grass-roots voters posed a formidable challenge to corrupt, corporate machine-controlled partisan politics in our state.

I will begin by briefly reviewing how Colorado women won the right to vote in 1893. The first women in the world to win full suffrage rights were in Wyoming territory in 1869. At that time, there were more antelope than people populating our neighbor to the north. Voting women in Wyoming became a national curiosity, with national magazines accusing them of being uncivilized and immoral champions of "free love." On the other hand, one New York magazine published images of well-dressed residents of Cheyenne registering voters in 1888, helping to dispel the notion that western women voters were "uncivilized" western frontierswomen.[2]

After women won suffrage in 1893, many served as ward workers canvassing and registering female voters in both working class and wealthy neighborhoods prior to every election. (From *The Circle Magazine*, October 1907). *Photograph courtesy, Denver Public Library, Western History Collection, #Z-8811.*

Colorado women attempted to win full suffrage with statehood in 1876, but fell short both in the constitutional convention, as well as in a public referendum in 1877. By the early 1890s, the population of women in the state had grown, and the political dynamics had changed. Suffrage leaders such as journalists Minnie Reynolds and Ellis Meredith joined with the leaders of the state's many women's clubs, such as Elizabeth Ensley, a black club leader in Denver. In 1890, they formed the Non-Partisan Colorado Equal Suffrage Association. The name reflected the suffrage leaders' political strategy of building a coalition that crossed party, class, racial, religious, and ethnic lines. The name also reflected the political climate of the time, in which not two but three major political parties held sway in our state: the Democrats, the Republicans, and the rapidly growing Populist or "People's" Party composed of farmers, urban workers, and unionized miners across the state.[3]

The suffrage cause was greatly aided by the remarkable election of Populist governor Davis H. Waite in 1892. Waite was an attorney and union leader in the Knights of Labor from Aspen. With dissatisfaction over Colorado's faltering economy running high, Waite was swept into office, together with a Populist majority in the state house. Women activists in the People's Party attended its founding convention in Omaha in July 1892, gaining the party's support for woman suffrage— then viewed as a radical reform. One of Waite's first acts as governor was to win passage of a bill to submit a constitutional amendment granting woman suffrage to the voters in a referendum election in November 1893.

As the suffrage campaign ensued during 1893, a disastrous economic depression hit Colorado, known as the panic of 1893. The federal government had rescinded the Sherman Silver Purchase Act, which ended government purchase of silver from Colorado to back federal currency. The state quickly shifted from a mining boom to a bust—throwing thousands of miners out of work, and into relief camps and soup lines in cities like Colorado Springs, Pueblo, and Denver. Many of the same women who were already active in the suffrage campaign began organizing relief efforts for the unemployed, homeless miners.

Angry, disgruntled, unemployed miners soon came to appreciate the women who were operating the relief camps. The politically savvy suffragists handed the men a cup of soup with one hand, and a suffrage leaflet in the other hand. The literature which the women distributed promised that if women won the vote, they would help reinstate the silver standard, "clean up" politics, and pass laws to benefit working people. Soon the majority of men around the state promised to register and vote for suffrage, declaring "Let the women vote! They can't do any worse than the men have."[4]

Suffrage organizers from all walks of life worked tirelessly to keep Colorado male voters sober on election day and to get them to the polls. Their efforts bore fruit on November 7, 1893, when the suffrage measure passed handily with a margin of over seven thousand votes. Newspaper publisher and Colorado suffragist Caroline Nichols Churchill announced the victory to the world in her women's weekly, *The Queen Bee*: "Western Women Wild With Joy Over Colorado's Election."[5]

The first four states in which women won full voting rights were all Rocky Mountain mining states: Wyoming (territorial 1869, state 1890), Colorado (1893), Utah (1896), and Idaho (1896). This indicates that organized miners may have been a key factor in supporting many radical or experimental political measures, including early women's suffrage. However, many political observers at the time were skeptical as to whether newly enfranchised women voters would help the labor movement, or side with the employers, at the ballot box.

In early 1894, Gov. Waite faced a formidable challenge in the midst of the deepening economic depression—a major strike in the Cripple Creek gold fields, led by the Free Coinage Miners Union No. 19, Western Federation of Miners. Their principal demand was for the eight-hour day, a basic standard for working people, which we take for granted today. The mine owners resisted the union's demands. In June 1894 Governor Davis H. Waite called in the state militia to keep the peace by protecting striking miners from gun-toting company guards.[6] The union then called on Waite to represent them in negotiations with the mine owners.

The resulting "Waite Agreement," included provisions for payment of $3 for an eight-hour day, as well as union recognition. The Waite Agreement provided the model for hundreds of subsequent union contracts in hard rock mining districts in the Rocky Mountain West over the next decade. Waite was at first credited with resolving the strike with a minimum of violence, but mine owners broke the truce despite the governor's efforts.[7]

Governor Davis Hanson Waite's pro-union "People's Party" also supported Colorado's victorious woman suffrage referendum in 1893. He also served as chief negotiator for the miners in the Cripple Creek strike of 1894, resulting in the model "Waite Agreement" granting a standard of $3 for an eight-hour day in the mines. *From* Representative Men of Colorado, *The Rowell Art Publishing Co., Denver, 1902, Special Collections, Pikes Peak Library District.*

The Western Federation of Miners (WFM), and many other unions across the state considered Davis Waite a hero.[8] But that did not help him win his bid for re-election in 1894. The Cripple Creek strike of 1894 had caused tremendous political repercussions statewide. The mine owners, who were principally represented by the Republican Party, organized a systematic anti-Waite campaign based on fear of union radicalism and promising a return to "law and order." This would not be the first time that events in the Cripple Creek gold fields influenced political events in Colorado.

In early 1894, Populist women, led by First Lady Celia Waite, organized Populist Women's Legions to register newly enfranchised women voters. Other women like Minnie Reynolds "stumped the state" for the People's Party as some of the first women in the country to perform as leading spokespersons for a political party. Large numbers of women registered to vote in Populist-leaning districts in mining areas including Teller, Fremont, and Huerfano counties.

But the People's Party was divided, and no match for the Republican Party political machine that operated principally in Arapahoe County (Denver). Newly enfranchised Republican women formed an "army of women" which rounded up thousands of new voters to cast their ballots against "Waiteism" and for "law and order." Voters in the election of 1894 delivered a stunning defeat for Gov. Waite, with Republican James McIntire taking over the governor's office in January, 1895.[9]

The Republican sweep also brought the historic election of the first three women (all Republicans) to the Colorado State House of Representatives: Clara Cressingham and Frances Klock of Denver, and Pueblo attorney Carrie Holly. Voters also elected a Republican as Colorado's first female State Superintendent of Public Instruction. Again, leaders of the labor movement wondered if the majority of newly enfranchised women would

Newly enfranchised women voters flocked to the polls in Colorado elections after 1893. Their votes often determined the outcome of state elections, including the highly contentious elections of 1894 and 1904. (From *The Circle Magazine*, October 1907). *Photograph courtesy, Denver Public Library, Western History Collection, #Z-8811.*

permanently back the more conservative Republican Party in future elections, making it even more difficult to gain laws and concessions which would improve the lives of families in mining districts like Cripple Creek.[10]

By 1896, women had established their "place" in the state legislature and began lobbying for laws to improve education, social conditions, and labor protections. For example, non-partisan candidate Martha Conine of Denver won a house seat in 1896, from which she advocated many reforms for working women, including the "laundry girls law," or the eight-hour day for women. Conine called on women's clubs, labor unions, and concerned citizens to support this measure to protect women from the adverse health and social effects of growing industrialism. A total of ten women won seats in the Colorado house by 1900, including several Populist and "fusion" candidates (combinations of several party factions).

Increasing numbers of women were employed in small factories, such as the Merchants Biscuit Company in Denver. The worst working conditions were in the industrialized steam laundries, such as the Western Steam Laundry in Denver's "red light district" (now the upscale district known as "LoDo"). Women worked in temperatures above 100 degrees Fahrenheit for ten to twelve hours per day without breaks. They were routinely exposed to toxic laundry detergents, and repetitive motion injuries from operating heavy steam presses and "mangles."[11]

Martha Conine and other female political leaders often spoke to crowds about the "laundry girls" law:

> There is no sense in saying that women want to work twelve and fourteen hours a day for small wages . . . for $5.00 a week. . . . In the interests of humanity, the public must interfere. . . . If we haven't the law, we have something higher—public opinion. . . . Only by appealing to the public can we aid in getting women workers what the good of humanity demands they have, the eight-hour day at manual labor.[12]

Due to the efforts of Conine, other middle and upper class women began to sympathize with the plight of working women, and working-class people in general. A group of female political activists formed the "women's auxiliary" of the Western Federation of Miners (WFM) in 1902, estimating that over two hundred thousand pro-union female voters could be mobilized to vote for pro-labor candidates in the Rocky Mountain states. Middle class and professional women formed a Colorado chapter of the Women's Trade Union Unity League, a national association which advocated unions and legal reforms for working women. During the late 1890s, the Colorado State Federation of Women's Clubs formed a "legislative committee" which served for two decades as a powerful lobbying group for

Bakery workers packing crackers at Merchants Biscuit Company in Denver were part of the female work force that stood to benefit from an eight-hour law for women in Colorado, passed by the voters in 1902. The state Supreme Court systematically overturned the law, as well as eight-hour laws for miners. *Photograph courtesy, Denver Public Library, Western History Collection, #X-23854.*

women's causes. These women's associations formed the basis for a growing alliance between women's interest groups and the mostly male union leaders and members.

Beginning in the 1880s, Colorado's male union leaders had worked for passage of an eight-hour-day law for miners and smelter men. To members of the union movement, the eight-hour day symbolized a basic respect for all wage workers. The issue provided a great rallying point for unions to unite other groups of fair-minded citizens against corporate owners who resisted any legal reforms which would cut into profits, such as minimum hours or wage laws. Despite a U.S. Supreme Court decision upholding Utah's eight-hour-day law for miners, Colorado's Supreme Court overturned a similarly worded 1899 Colorado law, declaring that it violated the "right of contract" between employers and employees.[13]

Concerned union leaders, women's organizations, and other progressive citizens groups organized a ballot initiative in 1902 to place eight-hour day laws for both miners and women on the ballot. The eight-hour measures both won handily. But according to Colorado law, a majority vote in the legislature approving the measures was required before the eight-hour laws could be enacted.

The Republican Party had out-campaigned the divided Populists and Democrats, resulting in the election of an extremist anti-union governor, James H. Peabody and a Republican-controlled legislature in 1902. The legislature failed to implement the vote of the people by failing to enact the eight-hour laws. This was the straw that broke the camel's back in most of the state's mining districts. Strikes broke out over local issues, but the eight-hour day remained the core demand in each dispute. The state had never been so politically polarized since Davis Waite's intervention in the Cripple Creek strike of 1894.

Under the militant leadership of William "Big Bill" Haywood and Charles Moyer of the WFM, a series of major strikes broke out in the mining districts of Colorado throughout the year of 1902–1903, including Telluride, Colorado City Smelter, several smeltering plants in Denver, and the Cripple Creek/Victor mining district. The United Mine Workers also launched major strikes in Colorado's southern coal fields.

Observers at the time, as well as many historians since, have concluded that the mine-owner-controlled Republican governor and legislature provoked the labor wars of 1902–1904. They had blatantly refused to implement a simple, basic demand that had the overwhelming support of Colorado voters—the eight-hour day. The state's political leaders also discouraged peaceful negotiations with the unions, in favor of breaking strikes by the use of force by military troops and police.

Women played an important role in supporting the miners in the Cripple Creek labor wars of 1903–1904. The Woman's Auxiliary of the Western Federation of Miners distributed relief supplies and funds. Journalist Emma Langdon of Cripple Creek documented the events of the strike, which she later published in a book under the auspices of the WFM.[14] Langdon claimed to be the first female member of the WFM.

Langdon and others documented countless examples of unconstitutional arrests, deportations, and an organized terror campaign launched by state troopers and local Teller County law enforcement officers against strike leaders, as well as rank and file miners. As we know, there was great loss of life in the ensuing violence and explosions, which remain unresolved to this day. Langdon played a heroic role by publishing the union newspaper after its entire editorial and production staff was arrested and placed in the infamous "bull pen" in 1903. When the dust of the strike settled, Langdon claimed that at least 42 men were killed, 112 wounded, 1,345 arrested and imprisoned, and 773 deported from Colorado.

In order to publicize the abuses of police power in the strike, the WFM circulated a now-famous poster throughout the state of Colorado and other Rocky Mountain mining districts. The poster prominently featured the stars and stripes of the American flag, listing blatant violations of the constitutional rights against the strikers on each stripe. The union wanted to contrast its own rights to free speech and economic justice with the "law and order" practices of Gov. Peabody, Attorney General Sherman Bell, and the Colorado Mine Owners' Association.

WFM President Charles Moyer was subsequently arrested and charged with "desecration of the flag" for circulating this poster. In response, J. C. Sullivan, president of the Colorado State Federation of Labor declared that: "Gubernatorial czarism has climbed to the loftiest summit of despotism: where military might rides rampant over the constitution and the law."[15]

Mary Harris "Mother" Jones. *From* The Cripple Creek Strike, *by Emma Langdon, Special Collections, Pikes Peak Library District.*

Another outspoken female critic of Peabody's approach to "law and order" was Mary Harris "Mother" Jones—a fiery organizer for the United Mine Workers of America. At age 73, Mother Jones was called to Colorado by the United Mine Workers to support striking families in the southern coal districts in 1903. While she was here, she called upon her friends in the WFM, and observed the events in Cripple Creek in her autobiography:

> The state of Colorado belonged not to a republic but to the Colorado Fuel and Iron Company, the Victor Company and their dependencies. The governor was their agent. Whenever the masters of the state told the governor to bark, he yelped for them like a mad hound . . . all civil law had broken down in the Cripple Creek strike. The militia under Colonel Verdeckberg said, "we are under orders only from God and Governor Peabody." Strikers were arrested for vagrancy and worked in chain gangs on the street under brutal soldiers. Men, women and tiny children were packed in the bullpen at Cripple Creek. Miners were shot dead as they slept. They were ridden from the country, their families knowing not where they had gone, or whether they lived. And why were these things done? Because a group of men had demanded an eight-hour day . . . that was all. Just because miners wanted a better chance for their children, more of the sunlight, more freedom. And for this they suffered one whole year and for this they died.[16]

Mother Jones had little patience for the women's suffrage movement. She feared that women's votes, especially those of the wealthy class, could hurt the working class and its unions. She declared, "I did not believe in women's rights nor in men's rights but in human rights. No matter what your fight, don't be ladylike!" But Mother Jones's and Emma Langdon's appeals to end these attacks on miners helped make an impact on many Colorado citizens, including women voters. Leaders of many middle and upper class women's groups felt that peaceful negotiations, not violent repression, would be the only way to resolve the issues.

The statewide political fallout from the Cripple Creek strike was tremendous, as revealed in the gubernatorial election in the fall of 1904. Not only were the striking unions outraged by Peabody and Bell's unconstitutional violations of civil rights and free speech, but large numbers of mainstream voters were outraged as well. The outcry of "anybody but Peabody" rose up from Denver to Durango, resulting in broad support for Democratic Party candidate Alva Adams (despite his spotty record on union rights). A strong backlash against "Peabodyism" developed among women voters in the state, led especially by Democratic Party activists Mary C. C. Bradford,

Martha Conine

Mary C. C. Bradford

Ellis Meredith

From *Representative Women of Colorado*, Williamson-Haffner Co., Denver, 1914.

Martha Conine, and Ellis Meredith. Democratic women's clubs joined WFM women's auxiliaries, women union leaders from the Colorado State Federation of Labor, and others to register and mobilize women voters.

"There is one class of voters in Victor and the Cripple Creek district who cannot be intimidated, and that is the women," observed a Denver reporter in 1904. "The women have registered as never before. They have gone to the registration places in flocks, and have passed the word along to all the women who were made widows by the deporting of their husbands . . . now is their chance to get out and rebuke Peabodyism."[17]

A woman delegate to the Democratic State Convention explained why the Cripple Creek women planned to flock the polls to vote for Alva Adams: "Our husbands are all deported, and we women have to defeat the Republican Party so they can come back again."[18] Typical of the "union widows" whose husbands had been deported was Mrs. Fred Minister, a Cripple

Creek laundress and mother of six, who warned anti-union politicians that: "We are all registered and we will vote!"[19]

Emma Langdon worked tirelessly for three months before the November 1904 election to register and mobilize voters against Peabodyism. She mounted podiums around the state, appealing to all who would listen about the attacks by Peabody upon the miners and their families in Cripple Creek. "Labor's victory awaits them at the ballot box . . . let organized and unorganized labor stand shoulder to shoulder and not be divided on election day," declared radical union journalist Emma Langdon.[20]

On election day, blatant voter fraud and violations of voting rights occurred across the state—especially in mining districts like Teller County. According to later testimony in the legislative hearings on voter fraud in the election, locally appointed poll watchers were ejected from polling places at the point of a gun. This included one woman poll watcher named Mrs. Kennedy in Victor, who was allegedly physically ejected from her post, and replaced by thugs hired by the mine owners. Gunmen threatened mine employees with their jobs on their way to the polls, and miners were allegedly promised a day's pay and a day off in exchange for their vote for the Republican Peabody. Similar tactics were used in the Telluride mining district, and in the southern counties of Huerfano and Conejos.[21]

Peabody contested the election of Alva Adams, despite the Democratic Party candidate's substantial margin of victory of over ten thousand votes. Republican Party leaders alleged major voting irregularities by the Democrats, especially in Denver, where Mayor Robert Speer controlled the infamous "Big Mitt" political machine.

The Republicans accused voters in working class, immigrant, and "red-light" vice districts of "repeating," selling their votes, and stuffing ballot boxes. Over twenty volumes of sworn testimony before the Colorado State House of Representatives were recorded in the hearings. Legislators claimed the evidence was so inconclusive that they could not determine who really won the election. This gave rise to the famous "three governors in a day" agreement, in which both Alva Adams and James Peabody held the office briefly, until the parties compromised on Republican Lieutenant Governor Jesse McDonald, who was sworn in—all on the same day.[22]

The extreme political polarization of the state in 1904, the campaign of fear of "terrorism," and the eventual "selection" of the governor in defiance of the decision of the voters are all eerily reminiscent of the campaigns and elections we are experiencing in our present time.

It was not until 1908 that a strong, pro-labor governor was finally elected in Colorado—John Franklin "Honest John" Shafroth. Under Shafroth, advocates of labor unions and working women enjoyed unprecedented political influence. During Shafroth's two terms, lawmakers enacted an array of labor protections, including a factory inspection act and laws protecting workers who joined unions. Electoral reform legislation, including measures providing for the initiative and referendum, the direct primary, and direct election of U.S. Senators also passed during Shafroth's tenure.[23]

Women's groups and labor unions united once again in 1912, when they successfully petitioned to place the eight-hour day for women and the eight-hour day for miners on the ballot as initiated measures. After decades of legislative delays and court reversals, voters approved these measures by decisive majorities. The progressive political coalition between workers, farmers, and middle class city dwellers had retained its strength for almost two decades since women won the vote in 1893. Colorado had become a national model for successfully mobilizing grass-roots voters from all walks of life, both men and women, to democratize their political system and their laws.

By the end of 1912, women in just nine western states had won the right to vote.[24] After two decades of suffrage, women political activists in Colorado had won themselves important places in partisan politics and legislative reform. They accomplished this in part through their own efforts, as well as by forming alliances with the leaders of Colorado's well-organized labor movement. The elections of 1894 and 1904 hinged upon the struggle between mine owners and miners in Cripple Creek. In both cases, women activists and voters had tipped the balance.

Notes

1. Emma F. Langdon, *The Cripple Creek Strike,1903–1904,* (Victor, Colorado: *Victor Daily Record,* 1904), reprinted as Emma F. Langdon, *The Cripple Creek Strike, A History of Industrial Wars in Colorado* (New York: Arno Press, 1969).

2. *Leslies' Illustrated Weekly Newspaper,* (November 1888). Image in the collections of the Colorado Historical Society.

3. A number of works on the Colorado suffrage campaigns exist, including Marcia T. Goldstein and Rosemary Fetter, "Let The Women Vote! Colorado Suffrage Centennial, 1893–1993," *Colorado Committee for Women's History,* 1993; Rebecca A. Hunt and Marcia T. Goldstein, "From Suffrage to Centennial: A Research Guide to Colorado and National Women's Suffrage Sources," *Colorado Heritage* (Spring 1993): 40–8; Billie Barnes Jensen, "Colorado Woman Suffrage Campaigns of the 1870's," *Journal of the West* 12, no. 2 (1973): 254–71, and "Let the Women Vote," *Colorado Magazine* 41, no. 1 (1964): 13–26; Stephen J. Leonard, "'Bristling for Their Rights': Colorado's Women and the Mandate of 1893," *Colorado Heritage* (Spring 1993): 7–13; Susan M. Marilley, "An Exceptional Victory: The Colorado Campaign of 1893," Ch. 5, *Woman Suffrage and the Origins of Liberal Feminism in the United States, 1820–1920* (Cambridge: Harvard University Press, 1996); Carolyn Stefanco, "Networking on the Frontier: The Colorado Woman's Suffrage Movement, 1876–1893," in the *Women's West,* eds. Susan Armitage and Elizabeth Jameson (Norman: University of Oklahoma Press, 1987): 265–76; and "Harvest of Discontent: The Depression of 1893 and the Women's Vote," *Colorado Heritage* (Spring 1993): 16–21.

4. *Rocky Mountain News,* January 1, 1894.

5. *The Queen Bee,* November 29, 1893.

6. For scholarly works on the 1894 Cripple Creek strike, see Elizabeth Jameson, *All That Glitters: Class, Conflict, and Community in Cripple Creek* (Urbana: University of Illinois Press, 1998); and George G. Suggs, *Colorado's War on Militant Unionism: James H. Peabody and the Western Federation of Miners* (Norman: University of Oklahoma Press, 1972, 1991).

7. Original handwritten Waite Agreement at Colorado State Archives, Davis H. Waite Papers, Box 27092, FF1. The full wording of the document follows:

> Agreement in re: Miners & Mine Owners, Cripple Creek, June 4, 1894; For the purpose of settling the serious differences between Employers and Employees in Cripple Creek Mining District El Paso County, Colorado—it is agreed by and between Governor

Davis H. Waite appointed by and representing the Free Coinage Miners Union No. 19 W.F.M.A. its members and other miners of said District, on the one part, and J. J. Hagerman and David H. Moffat, for themselves as mine owners and employers of mining labor in said District, on the other part—as follows:

1. That Eight hours of actual work shall constitute "a day", divided as follows: four hours continuous work—then twenty minutes for lunch—then four hours of continuous work, for which said eight hours of labor there shall be paid three dollars. ($3.00)
2. In the employment of men, there shall be no discrimination against union men or against non-union men.
3. The undersigned J. J. Hagerman and David H. Moffat earnestly urge upon other mine owners and employers of mining labor in said Cripple Creek Mining District to accede to and act upon the foregoing agreement.

Davis H. Waite [signature]
J. J. Hagerman [signature]
David H. Moffat [signature]
Dated June 4, 1894.

8. See, for example, a letter of support for Waite from the office of the Western Federation of Miners, Butte, Montana, October 10, 1894, which read in part, "To Members of the Western Federation of Miners and all Branches of Organized Labor in the State of Colorado: His [Waite's] action in the miners' strike at Cripple Creek, whereby he prevented a conflict which would have been most serious, saved the State from an outbreak which might not have been suppressed for years, should be enough to gain him the earnest support of every union man and every toiler in the State." Reprinted in J. W. Deane, "The Voice of Labor Preaching the Gospel of Human Liberty: Governor Waite, The Coming Emancipator of the Industrial White Slave," People's Party campaign brochure (Chicago, 1894), 3 (Denver Public Library, Western History Department).

9. For works on Davis H. Waite and the People's Party in Colorado, see John R. Morris, *Davis H. Waite, The Ideology of a Western Populist* (Washington, D.C.: University Press of America, 1982); James Edward Wright, *The Politics of Populism: Dissent in Colorado* (New Haven: Yale University Press, 1974); Marjorie Hornbein, "Davis Waite, Silver, and Populism," *Essays in Colorado History* (Denver: Colorado Historical Society, 1983). For an analysis of Waite's relationship with newly enfranchised women voters in the 1894 election, see John

R. Morris, "The Women and Governor Waite," *Colorado Magazine* (winter, 1967). A recent work which includes a discussion of Waite's tenuous relationship to the union movement in Colorado is John P. Enyeart, *"By Laws of Their Own Making": Political Culture and the Everyday Politics of the Mountain West Working Class, 1870–1917*, Ph.D. Thesis, University of Colorado, 2002.

10. My forthcoming Ph.D. dissertation, *"Meet Me at the Ballot Box,"* will analyze the role of newly enfranchised women in Colorado's political parties, and in non-partisan coalitions with other organized interest groups such as labor unions, during the initial years after they won suffrage in 1893.

11. Marcia T. Goldstein, "The Women's Eight-Hour Law in Colorado: Woman's Suffrage and Pro-Labor Reform in Progressive Era Colorado, 1893–1914," paper presented at the 5th Women's West Conference, Coalition for Western Women's History, July 27–29, 2000, Pullman, Washington.

12. Clipping contained in Martha Conine scrapbook (Denver Public Library, Western History Department). Conine served as a Colorado state legislator from 1897 to 1898, and was a leading activist in the legislative lobbying efforts of the Colorado State Federation of Women's Clubs until her untimely death in 1910.

13. David Lawrence Lonsdale, *The Movement for an Eight-hour Law in Colorado, 1893–1913* (Thesis, Ph.D., University of Colorado, 1963).

14. Langdon, *The Cripple Creek Strike.*

15. An original copy of the WFM "flag" poster is housed in the papers of Judge Benjamin B. Lindsey, Library of Congress. Quotation published in Harold Vincent Knight, *Working in Colorado: A Brief History of the Colorado Labor Movement* (Boulder: University of Colorado Center for Labor Education and Research, 1971).

16. Mary Harris Jones, *The Autobiography of Mother Jones*, edited by Mary Field Parton (Chicago: C. H. Kerr Pub. Co., 1990), reprint of 1925 edition, see especially Chapter XIII: "The Cripple Creek Strike," and Chapter XXII: "You Don't Need a Vote to Raise Hell."

17. Article by Del A. Matheis of the *Denver Times* October 25, 1904, quoted in *Miners' Magazine* 6:71 (1904), 7–10.

18. Helen L. Sumner, *Equal Suffrage: The Results of an Investigation in Colorado Made for the Collegiate Equal Suffrage League of New York State* (New York, London: Harper & Brothers, 1909), 6.

19. *Miners' Magazine* 6:71 (1904), 7–10.

20. Emma Langdon, *The Industrial Wars in Colorado* (1905), 161.

21. James H. Peabody, *Contestor, vs. Alva Adams, Contestee : Before the Joint Convention of the Fifteenth General Assembly of the State of Colorado* (Denver: State of Colorado, 1905), 14 volumes.

22. Marjorie Hornbein, "Three Governors in a Day," in Steve Grinstead and Ben Fogelberg, eds., *Western Voices: 125 years of Colorado Writing* (Golden, Colo.: Fulcrum, 2004).

23. Stephen J. Leonard, Thomas J. Noel, and Donald L. Walker, Jr., *Honest John Shafroth: A Colorado Reformer* (Denver: Colorado Historical Society, 2003).

24. Wyoming (1869, 1890); Colorado (1893); Idaho and Utah (1896); Washington (1910); California (1911); Oregon, Arizona, and Kansas (1912). For details on the suffrage campaigns in these states see "This Shall Be the Land for Women: The Struggle for Western Women's Suffrage, 1860–1920," online exhibit, Women of the West Museum, Autry National Center, http://www.autry-museum.org/explore/exhibits/suffrage/suff_time.html. See also Rebecca J. Mead, *How the Vote Was Won: Woman Suffrage in the Western United States, 1868–1914* (New York: New York University Press, 2004).

Is Colorado in America?: Emma Langdon, Polly Pry and the Colorado Labor Wars

Bridget Burke

The year 2004 marks the centennial of the Cripple Creek Strike. This conflict between miners and mine owners led to the declaration of a state of insurrection in Teller County, the military occupation of the Cripple Creek District, the suspension of civil rights, the removal of elected officials, and the forced deportation of union miners and their sympathizers. Over thirty people died in violence relating to the strike. Although the strike was initially about bread-and-butter issues of wages and hours, it quickly became a battle over the right of workers to organize and the right of owners to run their businesses as they saw fit. The civil, military and constitutional conflicts that came out of the strike established Colorado as the birthplace of western labor radicalism and reinforced the state's reputation as a safe haven for western business conservatism.

Booster literature from turn-of-the-century Cripple Creek promotes the richness of the ore deposits, but it also boasts about the community life of the district, describing not a rough gold camp but a thriving community with churches, schools, lodges and social organizations, cultural amenities, and a stable, law-abiding citizenry. In 1900, the district boasted over thirty thousand residents, making it the third largest urban area in the state. Unlike other mining camps with highly transient, mostly male populations, Cripple Creek residents included high percentages of women, families, and homeowners. These demographics suggest that the Cripple Creek strike was not just a labor conflict; it was about competing visions of community.

The strike was not an isolated event, but lay squarely in the middle of what contemporaries dubbed Colorado's "labor wars." The conflict was chronicled in the local, state, and national press, with partisan journalists delivering very different versions of the events as they played out. Two women journalists are the topic of this paper: Emma Langdon, a union printer and socialist, and Nell Anthony, a Denver reporter and socialite.

Emma Langdon's account of the Cripple Creek strike was first published in May 1904, and expanded and reprinted in May 1905 with the subtitle "A History of Industrial Wars in Colorado." She begins in no uncertain terms:

> The state of Colorado ceased under the administration of James H. Peabody, to be republican in its form of government, and became a military oligarchy. The expressed will of the people was ignored by their chosen representatives, thus bringing upon the state a series of calamities, the magnitude of which may now be readily seen.[1]

Emma Langdon, 1904. *From* The Cripple Creek Strike, *by Emma Langdon, Special Collections, Pikes Peak Library District.*

Emma Parker was born in 1875 in LaGrange, Tennessee. She married Charles Langdon, a linotype operator, and moved to Victor, Colorado, in 1903. Emma and Charles, and Charles' brother Frederick, worked at the *Victor Record*, a pro-labor daily. Langdon was a member of the Typographical Union and secretary of the District Trades Assembly and later of the Women's Auxiliary of the Colorado State Federation of Labor.

She earned the title "Labor's Joan of Arc" when, after the *Record* press was shut down by the military, and the operators sent to the bullpen, she returned to the *Record* office with a skeletal crew and helped to print and distribute the morning edition. In 1904 she left Victor, speaking around the state about military outrages in Telluride and Cripple Creek. She was made an honorary member of the Western Federation of Miners (WFM) at their 1904 convention, and an honorary member of District 25 of the Mine Workers when she visited the southern Colorado coal fields to investigate that strike. She attended the founding convention of the Industrial Workers of the World (Wobblies) in 1905 as an independent delegate, and was active in first the Democratic and, later, the Socialist Party.

Langdon published two books directly relating to the Cripple Creek strike, and two additional works on labor conflicts. She worked as a mediator and organizer for many years, and at her death in 1937 her portrait was published in both the *Denver Post* and the *Rocky Mountain News*, the latter also reporting that she "owned probably the largest collection of birds in the state." Langdon's daughter Lucile was awarded a $50 death benefit from the Denver Typographical Union, where Langdon maintained a life-long membership.[2] In her history of the strike, Langdon voiced the union viewpoint. But she was not the only journalist who became impassioned by the strikes that raged in Colorado at this time.

Nell Anthony. *Photograph courtesy of the Colorado Historical Society.*

Leonel "Nell" Ross Campbell Anthony was born in 1857 on a Mississippi plantation. After an early marriage, Anthony left her husband and moved to New York where she began her career in sensational journalism at the *New York World*. In the 1890s Anthony arrived in Denver and began to write for the *Denver Post* under the name "Polly Pry." While at the *Post* she covered a number of scandals, including the sexual abuse of girls at the State Home for Dependent Children. She also traveled, filing stories from South America and Europe. But the cause that gained her notoriety was her petitioning for the release of Alferd Packer, a Colorado mountain guide convicted of cannibalizing his clients while starving and stranded.[3]

Around 1902 Anthony left the *Post*, and in 1903 she began publication of *Polly Pry*, a weekly with the motto "to hold, as 'twere, the mirror up to nature," with the side line of "telling the truth and shaming the devil."[4] *Polly Pry* started out as a cosmopolitan gossip rag for Denver debutantes, but by the time it folded in 1904 it was largely devoted to labor issues and Denver civic government. Anthony was active in the Colorado commission for the 1904 St. Louis World's Fair and the Women's Press Club, and during the First World War she served in the Balkans for the Red Cross. She died in 1938 and is buried in Fairmont Cemetery in Denver, as is Emma Langdon.[5]

Langdon was a resident of Victor, and a participant in the strike in the Cripple Creek District. Anthony was a muckraking journalist and Denver doyenne who saw industrial conflict as the most important issue facing the state, claiming that "the Homestead Affair was a mere bagatelle compared" to Colorado's labor wars of 1903–1904.[6] While both women recognized the significance of the conflicts, there the agreement ends. Both covered the strikes in handcrafted publications that essentially bypassed the mainstream press. Both considered their accounts to be the definitive history of the Colorado labor wars, without ever recognizing the other's efforts. And both ultimately expanded their reporting on the conflict between labor and capital to a national scope.

This paper focuses on several episodes where the efforts of Langdon and Anthony either shifted, converged, or became the stuff of legend: *Polly Pry*'s shift from a gossip rag to an anti-labor weekly in the fall of 1903; Emma Langdon's efforts to publish the *Victor Record* the night of September 29, 1903; and the flag desecration suit brought over a WFM broadside that asked the question "Is Colorado in America?"

The Education of Polly Pry

Polly Pry began publication in September 1903, after Anthony left the *Post* when her editors asked for a retraction of an article she had written accusing a union leader in Telluride of advocating murder.[7] In introducing the first issue of her new weekly, Anthony addressed her readers, warning them that she fully intended to be opinionated: "I hope it will please you; if not, I hope it will seriously displease you, as in either event, the result will be the same: you will read the paper."[8]

The first issue promised revelations about Denver's debutantes, but the nature of those revelations was veiled: Denver's social set was rarely named by name, their follies not fully attributed. Anthony's coverage of the labor situation was not so coy. Early issues feature a poem mocking Sherman Bell, the military commander of the occupied Cripple Creek District, followed by an article on "what a queer lot of Adjutant Generals Colorado has managed to scare up."

Within weeks of beginning publication, there is a brief description of a visit to Cripple Creek, and several issues later a more impassioned attack on WFM agitators and praise for Governor James Peabody, who had sent in the military at the behest of the mine owners. From the beginning, Anthony claimed to support the right of labor to organize to improve their working and social conditions, while consistently demonizing union leaders rather than union men for the violence of the strikes in Colorado.

Perhaps Anthony, so successful in sensational exposés, got weary of reporting on Denver society. Most likely she simply could not resist the good story, high in human drama, which the strikes presented. The issue published just after Thanksgiving 1903 contained this manifesto:

> *Polly Pry* is the only paper in Colorado that dares print the truth about the strike situation, which is today the most serious thing that confronts all the people and interests of the state. . . . If you want to know the facts—whether you are a capitalist, a day-laborer, a union man, or simply one of the unfortunate individuals who must pay [severe] prices for the fuel to keep you warm [then Anthony would publish them].[9]

Polly Pry. Courtesy of the Colorado Historical Society.

Her conception of her readership was broad; it seems far-fetched that issues of *Polly Pry* would find their way to union hall reading rooms. Correspondence from readers found in Anthony's papers include letters from lawyers, businessmen, women reformers, and as we'll see later, Adjutant General Sherman Bell, who was a big fan. All praise her courage and honesty in investigating the strikes throughout the state.

Polly Pry initially focused its labor reporting on the effect of the coal strike on Denver consumers, as the price for fuel crept higher with winter approaching. Anthony traveled to Trinidad and the outlying coal towns and met with officials from Victor Fuel and Colorado Fuel & Iron, and with John Mitchell of the United Mine Workers. She traveled to Telluride and met with WFM officials, whom she described as "that cheerful aggregation of swaggering ignorance and violence." The next day, December 5, just a week before the Cripple Creek District was declared to be in a state of insurrection, she "[made] her manners" to the "circle of bomb-throwing, train-wrecking, infernal-machine-sending cutthroats and assassins who are said to be the real leaders of the same Federation in Cripple Creek."[10]

If Anthony truly believed this, she was a brave woman indeed for having tea with them. Closer to home, Anthony recounted the Denver "food strike" of May 1903 which involved retail clerks, restaurant workers, and teamsters, and the Denver smelter strikes. Within a few months of beginning publication, labor news was taking over *Polly Pry*, which was now running 16 to 20 pages with a full half of it dedicated to "industrial conflict."

Social gossip had receded into the back pages. Anthony recognized this shift when, in early January 1904, she inaugurated the first in what would be a nine-part series on the "story of organized labor in Colorado in the fight of 1903–1904."[11] Later, *Polly Pry* included reports from correspondents in San Francisco, Chicago, New York and Indianapolis, on strikes from delivery boys to hearse drivers. She promised, "wherever there is a strike, our man will be on the spot."[12]

The *Victor Record*

While Nell Anthony was making forays onto the battlefields of Colorado's labor wars, Emma Langdon was living in the militarized zone, in Victor in the Cripple Creek District, working as a printer at the *Victor Record.* In August 1903 the *Record* became the official organ of the WFM, and at that time it had the largest circulation of any paper in the district. After the paper ran an uncomplimentary report on a member of the National Guard, soldiers stormed the *Record* office on the night of September 29, 1903. The printing force was marched off to the bullpen and held without charge for several days.

Emma Langdon had been active in the district's labor community since her arrival in Victor, but she gained celebrity for the events of September 29 and 30. As reported in an official statement of the WFM:

> [T]here can be very little chance of the strike being broken when the women are willing to take their husbands places when they become the victims of these blatant military despots. This happened last night when Mr. Charles G. Langdon was unlawfully thrown into the bullpen and Mrs. Langdon nobly took his place as the operator of a linotype machine at the *Record* office. . . . She has been working all last night and today to enable this issue of the *Record* to be gotten out. It is needless to say that the union men are all loud in their praises of this modern Joan of Arc.[13]

The report does not identify Langdon as an active unionist, or a competent journeyman printer, but as a spouse. Her role is described in terms of marital helpmeet. Langdon's story, that she left her infant daughter with a neighbor, ran to the *Record* office in the middle of the night, still in slippers and bedclothes, set the issue, went home to change, and then hand-delivered the September 30 newspaper to her husband and others imprisoned in the bullpen, has been both enlarged and disputed. The *Appeal to Reason*, a socialist paper, described the scene this way:

> One night, the soldiers swooped down on the plant and carried off the entire office force to the bullpen. Then it

> was that a plucky little woman, Mrs. Emma F. Langdon, wife of one of the Linotype operators, proved equal to the occasion, and, mustering a corps of women, finished the setting and printing of the paper.[14]

According to Langdon's own account it was A. Q. Miller, the business manager, and a pressman named Conrad who were in the office with her that night, and in a later recollection Miller claimed that he had composed the bulk of the paper and the headline "Somewhat Disfigured but Still in the Ring."[15]

Langdon's actions received a full page spread in the *Miners' Magazine*, under the title "A Colorado Heroine." The *Magazine* reported that Langdon was given a gold medal and noted that "the brutal henchmen of the Mine Owners' Association . . . acted without reckoning on Colorado women. . . . The men on strike . . . look with reverence and gratitude on the shrinking little woman who dared defy the mandates" of the military.[16] The piece was accompanied by a portrait of Langdon and a poem of several stanzas by Ida Crouch Hazlett. Frank Mannix, Teller County Clerk and a fellow member of the Typographical Union, presented the medal, voicing his belief that Langdon's work would "have a national scope."[17]

Langdon, much more so than Anthony, reminded her readers of her sex, her Southern heritage and physical frailty, and in that sense the accounts that emphasize her role as helpmeet above her status as paid journeyman printer may only reflect what she herself believed.

Langdon was active in the Victor Women's Auxiliary of the Colorado State Federation of Labor, formed just two weeks after the arrest of *Record* printers. The *Record* reported, "Excelsior Engineer's Hall on Victor Avenue was literally crowded with feminine beauty . . . the same being the wives, mothers, sisters and sweethearts of the union men of the district." Langdon was elected secretary of that group, which was generally charged with promoting the cause of unionism through supporting union-made goods and union-friendly stores, and "incidentally, to assist in putting the union's enemies out of business and winning the strike."[18] In the summer of 1904 the Women's Auxiliary would be accused of fabricating stories about the "whitecapping," "sweating," and deportation of union sympathizers.[19]

In closing the first volume of her history of the strike in the spring of 1904, Langdon evoked her status as a woman: "I have a few apologies to offer, not for anything I have said . . . but for the appearance of this work. It is the work of one woman, only an apprentice at the linotype." She describes her involvement in every stage of production, while also "doing [her] own sewing, baking, washing and ironing and other work that falls to the lot of women . . ." at the same time continuing to work for wages as a printer. Langdon asks for charity "in criticism of the mechanical part," but makes no apologies for the content of her history.[20]

Langdon attended the WFM convention in Denver in June 1904, and was away from the district the day of the Independence Depot explosion, in which thirteen non-union men were killed. She was advised by the military not to return to Victor. She remained an advocate in exile, but her activism was politicized by the failure at Cripple Creek, as demonstrated in her canvassing for the Democrats in the 1904 election. For Langdon, the failed strike demonstrated "how impotent are the efforts of organized labor in gaining better conditions when relying solely upon the strike in a contest—it should teach the wage earners the necessity of not alone depending on organization in the industrial field but of active participation in the political field."[21]

In the meantime, Nell Anthony was advertising in *Polly Pry* for her "History of Labor Troubles in Colorado from the start to the present time—the only history of the kind printed anywhere."[22] Her weekly had published exposés accusing Mother Jones of being a prostitute, and ran a column entitled "*Polly Pry*'s Role of Honor" which exposed union organizers, primarily WFM men, as murderers, convicts, anarchists, and dynamiters.

Praise for Anthony came not from the labor community, but from none other than Sherman Bell, the military commander in the district. Despite being ridiculed in *Polly Pry* (and every other newspaper: Bell was the buffoon, the comic relief of the strike), the adjutant general sent Nell Anthony a three-page letter praising the May 28, 1904, issue, declaring *Polly Pry* "the best publication in the world today."[23] Anthony had reported on Bell's intended resignation from the National Guard, and featured a full page signed portrait of the soon-to-be-former adjutant general. After the Independence Depot explosion, a state of insurrection was again declared in Teller County, and Bell's term was extended.

Line of horse-drawn hearses move down Bennett Avenue past lines of onlookers. Identified as "Bennett Avenue from Colorado Midland Terminal; hearses carry victims of Independence Depot bombing." *Photograph from Special Collections, Pikes Peak Library District.*

Is Colorado In America?

After the first attempt at military suppression of the *Victor Record,* the editors noted wryly "the *Record* does not know at this time how far it will be allowed to express its opinion under the reign of military law. But we propose to proceed just as if we lived in Free America."[24]

The year 1904 was the 100th anniversary of the Lewis and Clark expedition and of the Louisiana Purchase. This anniversary was marked by the St. Louis World's Fair. The question "Is Colorado in America?" was particularly resonant during this year of national celebration. Not unlike the current bumper sticker that reads "Hey dude, where's my country?," the WFM broadside circulated during this time challenged Coloradoans to consider the events unfolding in light of American traditions of democracy and freedom.

The poster featured an American flag overlaid with a list of military abuses, and the image of a Telluride striker handcuffed to a telephone pole. When Haywood and Moyer asked the question "Is Colorado in America?" they based their query on the assumption of basic rights—free speech, free press, the right to bear arms, trial by jury, arrest or search by warrant—that were both legally and collectively embraced as core American values.

Anthony asked the same question, with American freedom defined in capitalist terms. When non-union miners are under armed guard for their protection, she asks, "And this in Free Colorado?" Would "the affairs of the United States be controlled by Americans or by labor unions," she asks in February 1904. When detailing the harassment of scabs in Cripple Creek, the shooting of a mine manager in Telluride, or the adoption of the socialist platform by the WFM, "And this in Free Colorado?" becomes her refrain.[25] Similarly, Langdon follows every account of military abuse, every example of civil rights trampled, with the same question. It became the battle cry for both sides, the measure against which all actions were scrutinized.

Anthony was instrumental in instigating the flag desecration suit against Charles Moyer, president of the WFM, for distribution of the poster. In March 1904 Anthony printed a call to action, declaring that "the Sons of the [American] Revolution and other patriotic societies have a mission to perform in putting a stop to the recent desecrations of the American Flag by the nest of anarchists in the Mining Exchange Building . . . [who have turned] another trick by painting their foul lies on the flag of our country."[26] She was soon pleased to note that the Sons had "promptly obtained one of the circulars and appointed a committee to . . . file a complaint against Moyer and Haywood for desecrating the flag."[27] The Denver chapter of the Daughters of the American Revolution was organized around the issue, and met with what Anthony called the "hobby of nothing less than the punishment of those who desecrate the American flag."[28]

This poster, which exists in at least three versions, also provoked Langdon. In articulating the need for expanding

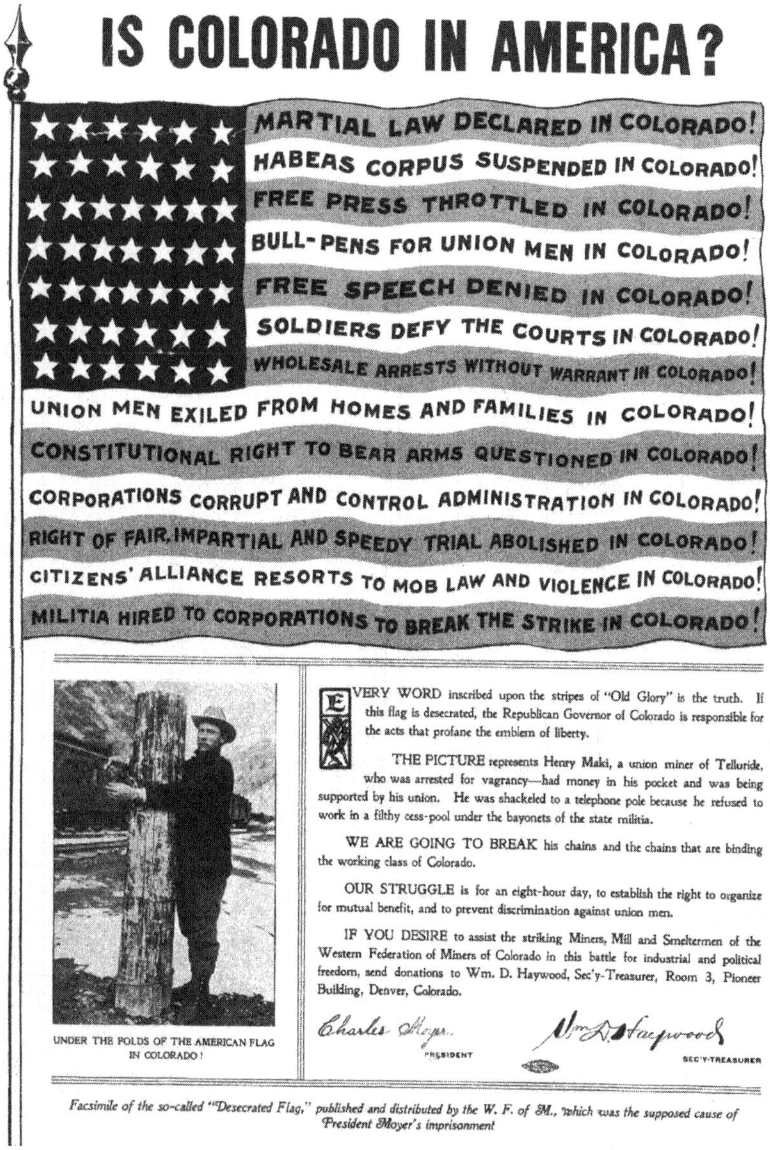

WFM poster with image of Henry Maki, 1904. *Courtesy, Denver Public Library, Western History Collection.*

her book, Langdon notes, "at [the end of part one] I felt that nothing could take place that would surpass the outrages that had been perpetrated on the citizens of Colorado." She then raises the case of Telluride miner Henry Maki, the poster boy of the WFM broadside. Langdon relates how Maki was "chained to a telephone pole in a bitter snow storm until nearly frozen"[29] for refusing to work as part of a crew of imprisoned "vagrants." Langdon was incredulous that, even after this "outrage . . . many people, who like to look on the bright side of things still maintained that Colorado was in America." Regarding the propaganda value of the image of Maki handcuffed to a telephone pole, Anthony was caustic as usual:

> The intense indignation of the people of this state over the shackling of a telegraph pole to one of the vagrant strikers of Telluride has subsided. The wave of popular disapproval of the act fairly swept the state. In Denver citizens organized and armed themselves to go to Telluride and free the telegraph pole.[30]

The flag suit, once in court, was quickly settled, though not at all to Anthony's satisfaction. She voiced her disgust with the patriotic clubs, who must surely have included among their members the debutantes and Denver elite that were her original audience:

> Perhaps some day when the Daughters of the [American] Revolution are quite done telling each other how distinguished their several ancestors were, they'll do something about this flag law. . . . Maybe, sometime in the dim and distant future when the Colonial Dames grow tired of bragging of their fathers . . . there will be a law passed in the state that will make it possible for a Colorado citizen to show his honor for the flag. . . . Mr. Moyer, the flag desecrater, could not be arrested because the laws of Colorado do not call the use of the flag as an advertisement a desecration.[31]

At the trial, Bill Haywood, WFM secretary, displayed consumer products bearing images of the American flag, including cigar boxes, tomato cans, and the business card of the Pinkerton Detective Agency.

While Moyer and the WFM got off on a technicality, calling the broadside an advertisement for the WFM, Langdon addressed the issue of what had really been "desecrated."

> Is Colorado in America? If you consult the map you will find it there. If you read the facts in this recent industrial struggle in the light of American history and traditions you will find nothing to recall the memories of our country's youth. . . . Desecration of the flag? Was it not the deeds done under it and not the truths inscribed upon it that constituted the desecration?[32]

Within the district, the same sentiments were put forth by residents who objected to the army in their midst. The citizens of Teller County petitioned the governor to remove the troops, stating that their presence was "unwarranted by the facts and . . . anarchistic in conception, spirit and consequence . . . [and] an invasion of our rights and liberties."

A telegram to Moyer, signed the "Victims of Military Despotism," accused the "powers that be [of] using the glorious American flag to cover crimes against the constitution of the illustrious state of Colorado, and the Bill of Rights . . ." They implored Moyer to "please arouse the citizens of this state and save us from anarchistic anarchism, militarism, un-Americanism, blatherskite-ism, and Bellism." The petitioners were among those who, in Langdon's words, "still . . . believed they were in Free America."[33] But with the suspension of *habeas corpus* under military law, she was asking "In what country do we live?"[34]

The question "Is Colorado in America?" was raised one more time in *Polly Pry*. A cartoon from the cover of the June 1904 issue shows the deportation of strikers out of state. Ironically, the image has Colorado in America but on the wrong side of Kansas. An error by the engraver, but one that reflects the disorientation engendered by the conflict.

Conclusion

In the end, Cripple Creek was about community—whose vision of community would prevail. Two women shaped that vision, both as journalists working outside of the mainstream press, essentially self-publishing their work. Both launched their careers and began their life-long passions for, or against, unions during the strike, and both achieved a national reputation and audience for their work. Their reporting, in its totality, is a seemingly strange mix of the distinctly feminine—gossip columns and patterns for crocheted reticules—and the arena of politics and war.

During the summer of 1904, all pretensions of civil law ceased to exist in the Cripple Creek District. Anthony could report in the June 11 issue that: "[C]all it mob law if you will, but it was a mob actuated by the highest civic motive. They took matters into their own hands, and now there is not a WFM member in the district who is not in prison, and there is not a union sympathizer holding office in the county."[35]

For Anthony, the persistence of community action, even outside the bounds of legitimate civic government, offered the only hope against anarchism, violence, and Western Federation rule.

In closing the expanded edition of her history of the strike, Emma Langdon finds the "bright side" in the persistence of place and in the persistence of a people dedicated to republican principles. In the end, landscape and community persevere, and therein lies hope. She states,

> [just] because we had a political accident as governor . . . and he appointed a military crazed adjutant general, is no proof that Colorado is not one of the greatest states in the union. Suppose we did have a Peabody and a Bell. There are no more like them and we have repudiated them. . . . Thank God Colorado is greater than Peabody . . . we love dear old Colorado.[36]

The Colorado labor wars were hard-fought, hard-lost, horrific and heroic; but not unique in either impetus or outcome. Langdon was correct in reminding her readers that industrial conflict was not new—that Colorado was not the only "war zone" on the map, that labor conflict was a "world problem."

In voicing a love for her adopted western landscape, Langdon ends her history of the Cripple Creek strike with the question that so engaged and impassioned Nell Anthony, as well as unionists, mine owners, politicians, the courts and the militia: "Is Colorado in America? Yes indeed, my dear reader. It is also on the map of the world."[37]

◊◊◊◊

Notes

1. Emma Langdon, *The Cripple Creek Strike 1903–1904* (Victor: *Victor Record*, 1904).

2. Gary Fink, ed. *Biographical Dictionary of American Labor* (Westport: Greenwood, 1985); Hurwitz Papers, Western History and Genealogy Department, Denver Public Library; International Typographical Union Papers, University of Colorado at Boulder; *Denver Post* and *Rocky Mountain News* obituaries, December 1, 1937.

3. Modupe Labode, "Polly Pry," *Colorado Historical Society Sandwiching in History* series, March 12, 2003.

4. *Polly Pry*, Vol. 1 no. 1, September 5, 1903.

5. Biographical Note, Finding Aid to the Leonel Ross Anthony O'Bryan (Polly Pry) Papers, Western History and Genealogy Department, Denver Public Library; Labode.

6. *Polly Pry*, Vol. 2 no. 2, March 12, 1904.

7. Labode, quoting Isabel Ross, *Ladies of the Press* (New York: Harper, 1936), 564.

8. *Polly Pry*, September 5, 1903.

9. *Polly Pry*, November 28, 1903.

10. *Polly Pry*, December 5, 1903.

11. *Polly Pry*, January 2, 1904.

12. *Polly Pry*, January 30, 1904.

13. Official Statement of Union District No. 1 Executive Committee, WFM, quoted in Langdon, 1904, 156.

14. Allan W. Ricker, *Appeal to Reason*, January 30, 1904.

15. Edith Stuart Jackson, *The History of Journalism at Victor, Colorado, in the Cripple Creek Gold Mining District* (Thesis, Colorado State College of Education, 1958), 121.

16. *Miners' Magazine*, 10–11.

17. Langdon *1904*, 231.

18. *Victor Daily Record*, October 15, 1903.

19. Emma Langdon, *The Cripple Creek Strike, A History of Industrial Wars in Colorado 1903–4–5; Being a Complete and Concise History of*

the Efforts of Organized Capital to Crush Unionism (Denver: Western Publishing Co., 1905), 330.

20. Langdon *1904*, 247.
21. Langdon *1905*, 270.
22. *Polly Pry*, July 16, 1904.
23. Anthony Papers, Correspondence.
24. *Victor Daily Record*, December 5, 1903, 1.
25. *Polly Pry*, January 16, 1904; February 6, 1904.
26. *Polly Pry*, March 5, 1904.
27. *Polly Pry*, March 11, 1904.
28. Ibid.
29. Langdon *1905*, 260.
30. *Polly Pry*, March 19, 1904.
31. Ibid.
32. Langdon *1905*, 297.
33. Langdon *1904*, 108–109.
34. Langdon *1904*, 137.
35. *Polly Pry*, June 11, 1904.
36. Langdon *1905*, 454.
37. Langdon *1905*, 455.

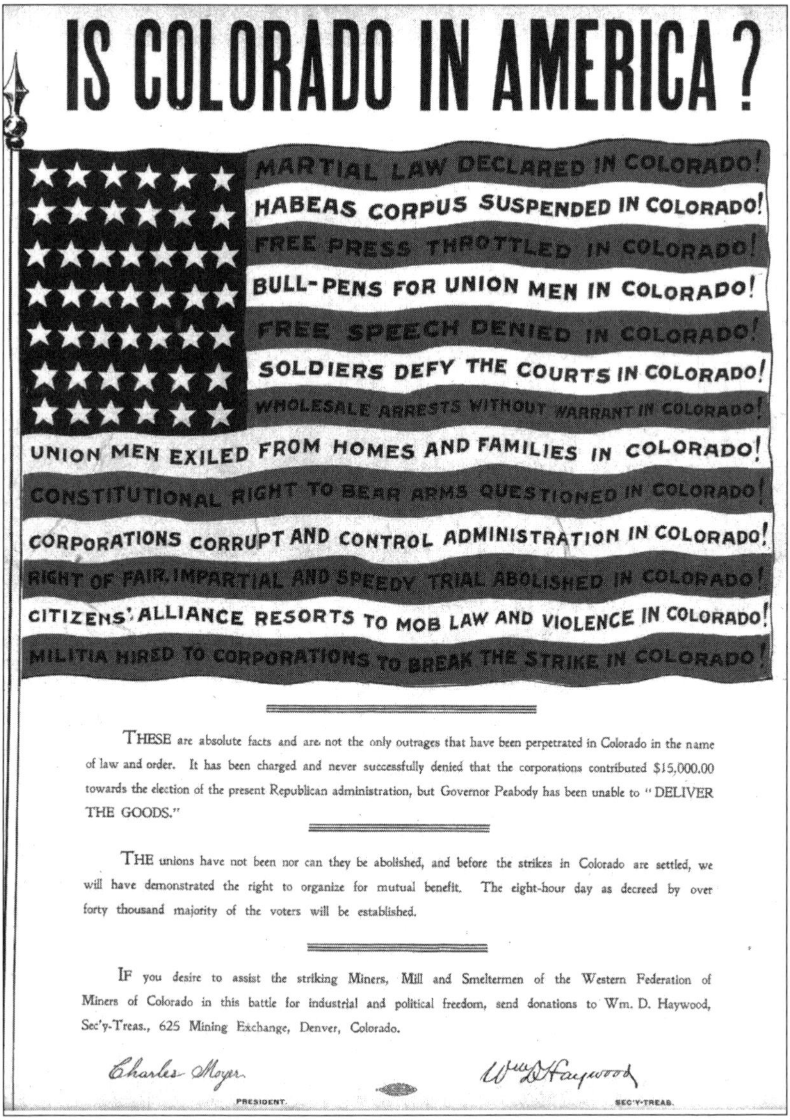

This incarnation of the "Is Colorado in America?" poster turned up in an ironic place during research—Governor Peabody's scrapbook. There is now a full-sized color photocopy framed in my office, next to the "mug shot" of John Harper. The colored flag portion remained the same in both versions, and appears on the cover of this volume. Note, however, that the language underneath, as compared to the one on page 64, makes clear that this one was earlier. The version on page 64 introduces Henry Maki, Telluride, and the accusations of "flag desecration." — Katherine Scott Sturdevant. *Courtesy of James H. Peabody Collection, SMss, Archives, University of Colorado at Boulder Libraries.*

Prostitution During the Cripple Creek District Labor Wars, 1904

Jan MacKell

There is an old western proverb that goes,

> The miner came in '59
> The prostitute in '61
> And between the two,
> They made the native son!

I think that little ditty was even truer than one might think, since prostitution was a major industry during the settlement of Colorado, and a presence in every boomtown across the state. In a place where the chief industry was mining, men could outnumber women as much as seventy to one. Often miners and prospectors would migrate from far away places and settle in a town or camp until they could afford to send for their families. The process could take months, or even years, during which lonely men yearned for the sight of a woman—any woman. Thus red-light districts and brothels were highly acceptable among male-dominated communities, including company towns and especially in mining towns.

The city of Victor was no exception to the rule. Evidence of Victor's illicit nightlife was already blatantly present during labor strikes that took place in the Cripple Creek District during 1893 and 1894. During the strikes, Mrs. Haillie Miller, a local prostitute, was employed by mine owners to bed striking miners and obtain information from them. It was a very limited success. Haillie was often far more interested in the whiskey before her than listening to an unhappy miner rave about his job.

Despite that failure, prostitutes played an important role in the life of both miners and mine owners. Most brothels served not just as a place to procure sex or play a little poker, but they also offered security and discretion. Men could talk more freely in a bordello than they could in the bars and businesses or at work or even at home, and the soiled doves kept their secrets.

As a result, these women earned the respect of their clients. It was a nice exchange: the girls were able to make their living, and the miners found a place to relax, enjoy themselves and get their minds off of their dangerous working conditions.

Poker and card games were other pleasant diversions offered in local brothels, where men felt free to speak out about their working conditions. *Photograph from Special Collections, Pikes Peak Library District.*

Many prostitutes made loyal friends out of their favorite customers, a varied lot from all walks of life. A good many of them were miners and young single men, but they could also be millionaires, business owners, laborers, city officials, and even law enforcement officers, husbands, and fathers. At the same time, many prostitutes in the district were prominent madams with extensive holdings at the local bank. Many a mine owner's indiscretions were never divulged to the public—saved by those working girls who zipped their lips.

Many prostitutes also invested in mining stocks, which was very fashionable for the day, much like owning the latest dressware or having a poodle. It was a status symbol, and naturally the gals who bought stocks in the mines supported both the mine owners and their employees.

In fact, many girls rented from, or catered to, the millionaire mine owners around the district, almost serving as double agents much like Mata Hari did during World War I. But they could be bought. One of my favorite stories is the time an associate of Spencer Penrose was sent to collect the rent from Cripple Creek madam Lola Livingston. Forty minutes later the red-faced associate returned without the money, with the explanation that he and Lola had decided he should take the debt out in trade.

By about 1895, the district millionaires were finding it necessary to protect their class status. Prominent men like Penrose, Charles Tutt and Winfield Scott Stratton were forced by their wealth to quit making appearances on Myers Avenue in Cripple Creek and the red-light district in Victor in order to save their reputations. But making gobs of money failed to stop these men from their love of working girls.

"Spec" Penrose in particular seems to have gotten into trouble with various girls, at least until his new wife Julie straightened him out. There was a widely circulated story about Penrose's involvement with a horse trainer named Sally Halthusen. Sally wasn't exactly a prostitute, but she was looking for a rich husband. It was said Sally had already been paid off once by the father of a man she was looking to marry. Eventually Spec's brother talked some sense into him and he stopped seeing Sally, at least in the public eye.

Winfield Scott Stratton.
Photograph from Special Collections, Pikes Peak Library District.

Incidents like this were enough to make the millionaires of Cripple Creek hire private call girls to come to their homes. Still, embarrassing incidents were known to rear their ugly heads now and then. One call girl, Candace Root, sued Stratton in 1895 for $200,000 with the charge of breach of promise to marry. Root claimed Stratton had lured her to his bed and also that she was pregnant. The case was dismissed.

Penrose and Tutt, who owned a racetrack and casino in Gillette and several brothels and "theaters" along Myers Avenue, were also fired on. The *Cripple Creek Morning Journal* called their Topic Theater a "variety den and skin dive" and said their dance hall should "be removed with the rest of the filth."

By the time Victor was officially founded in 1893, the city was chock full of saloons and whorehouses. Most of them were along portions of Third and Fourth Streets, but unlike most larger cities, Victor's red-light houses were scattered all over town. Victor, with its large male population of miners, was also unique in that brothels seemed welcome to coexist with other businesses throughout the town—at least for a little while.

In 1896, the miners in Victor got a taste of what was coming when they heard about labor strikes going on in Leadville. One of the stories that made its way to Victor was about Laura Evens, later a prominent madam in Salida, who took part in the labor war herself. Union men were blocking the entrance to one of the mines, much like they did at the town of Altman in the Cripple Creek District during the 1894 strike. No one was allowed to leave. It was Laura who showed up under the guise of visiting a friend who had not been allowed to leave, and was permitted to enter. What the guards didn't know was that she was smuggling the payroll for non-union miners under her skirts. Her effort was rewarded by a dinner invitation to the mine owner's home plus $100.

By then the three hundred block of Victor Avenue and all of Third Street fairly seethed with taverns, gambling houses and bordellos. Prostitution was definitely a part of everyday life. In time, however, Victor's proper women began objecting to the number of prostitutes plying their trade in town. Victor began regulating the red-light industry with fines and required health exams. Then Victor suffered its infamous fire in 1899, which burned down much of the town. When it was discovered the fire began behind a dance hall on south Third Street, Victor's prostitutes were even more ostracized.

By the time the labor wars began in 1903, Victor's red-light ladies had lost much respect. A good number of them were siding with the striking miners, which of course angered the mine owners. At the time, however, there were at least a handful of girls who were willing to spill their guts for a certain price. This labor war was definitely more violent than the strikes of 1893, and even Victor's prostitutes fell victim to the turmoil.

One of the most devastating incidents to take place happened on September 1, 1903. At that time, and for some years before, the newspapers had been making light of the plight of prostitutes who fell victim to abuse, drug addiction, and alcohol. A pro-union miner named Slim Campbell had been arrested for striking, but was released. Like thousands of other miners, Slim was frustrated and angry. Later that evening, Slim procured the services of a working girl and brutally murdered her. The mine owners blamed the Western Federation of Miners for allowing Slim to escape, even as they themselves continued to condemn Victor's prostitutes for offering refuge to men on both sides of the strike.

A few men, such as world traveler and author Lowell Thomas, defended Victor's bad girls. After all, they too were caught up in this bloody and heart-wrenching war. Thomas delivered newspapers in Victor as a boy, and his route included several brothels. Thomas described the ladies who, he said,

> came to their doors in their wrappers, and sometimes less, and appeared relieved to find only the newsboy. They often chatted with me and I answered respectfully, as I had been taught to do, and this seemed to please them inordinately. . . . After learning about what went on behind those shuttered windows I remained respectful to the girls, and they were always nice to me. When I finally got my burro I promptly led it over and showed it to them. They seemed proud, too.

Not surprisingly, it was a woman of questionable character who became the first female ordered to leave the district. On about June 25, 1904, just a few weeks after Harry Orchard blew up the depot at Independence, Mamie McGorraity of Victor got drunk and began traipsing up and down Victor Avenue. Mamie was arrested when she greeted a soldier with profanities, followed by her saying, "You scab herder." Mamie was arrested and taken to city jail. Upon her release she was told to leave the district, but the paper did not report whether or not she did so.

Like everyone else in the district, Victor's prostitutes waited out the strikes to their end, and it worked. In at least one way, they were very much like their male counterparts in the community: they were just trying to make a living and earn a little cash in order to survive.

Crapper Jack's saloon and dance hall, Cripple Creek, Colorado. *Photograph by Sanborn, from Special Collections, Pikes Peak Library District.*

For more detailed information about this topic, see Jan MacKell's book, *Brothels, Bordellos and Bad Girls: Prostitution in Colorado 1860-1930* (University of New Mexico Press, 2004).

High-Grading: Theft or Wage Supplement Right?

Ed Hunter

Mr. Daniel Webster defines high-grade as "being of superior quality." To underground copper miners, high-grade was dynamite containing more than 40 percent nitroglycerine, but to gold miners it means ore with high gold values.

The act of high-grading to the gold miner was removing, for himself, rocks from the mine containing high values of gold which he thought was only fair for he was the one making the mine owner rich. His union leaders told him that "wealth belongs to the one who creates it," but the mine owner considered this as theft. Gold ore could run many dollars per pound in some of the richer mines.

Now, if I've just blasted, and then pick up a shiny rock from the muck pile and stick it in my pocket to take home to show the wife and kids, is it stealing? If I go into the mine office after hours and remove a gold brick, is that stealing? I think we'd all agree on the latter case, but where the line was drawn was a matter of perception. Thomas A. Rickard, Colorado State Geologist, wrote that high-grading occurred when "greed overcame conscience."

High-grading of gold generally can occur only with deposits that have very high values in gold and there is a place to sell it. Did high-grading occur in the Cripple Creek/Victor Mining District? You bet! For the first 10–15 years gold ore values per ton were very high. Rather ironic when it took so many years to realize that this area even contained gold. At the turn of the century, a miner got $3 per day for his labor. If he were able to bring out some of the gold for himself, was it worth the effort? Again, you bet!

At the El Paso Mine, Calaverite crystals (24 percent gold) were found in locally named "bug" holes. A drill hole into one of these cavities would produce cuttings that averaged $14 per ounce. A miner could put some drill oil or candle wax in his hair, add the drill cuttings and go home with close to an ounce

The Jewelry Stope on the second level of the El Paso produced over $5,000,000 in gold, truly a high-grader's paradise. *Photograph courtesy of Ed Hunter.*

of gold. Even though he could only sell it for half the going price of $20.67 per ounce, the miner sometimes doubled his day's pay.

Stratton's great Independence Mine in the early years had values that averaged 4.6 ounces per ton (valued at over $1,800 on today's market). Picking out and removing a few spectacular specimens from the blasted muck pile could net the miner an ounce or two a day, again doubling his day's pay.

Burns, Doyle and Harnan, Portland Mine locators, secretly hand carried bags of high-grade down the mountain at night for shipment to a smelter. This was to build a defense fund to fight inevitable apex suits. They grossed almost $300 per night per man for three months. This was truly a high-grade mining district.

But this high-grade gold was down in the mine, so how was a miner to get it out of the mine knowing that the owner's representative would be at the surface on the lookout for such practice? The metal lunch bucket was ideal, but the handle had to be augmented with a shoulder strap as the regular handle might pull away from the bucket with a heavy load. A coat, shirt or pants liner with inside pockets was helpful. Hats were modified to have extra pockets, as were high-topped boots. Even the handle of the iron candleholder used for lighting underground could be enlarged to hold fine gold material in the curved handle which, when smeared with clay, could escape casual scrutiny.

Underground, a piece of pipe with a solid bottom was used as a pulverizing mill, reducing a rock in the pipe to powder by pounding it with a drill steel so it would pack into the candleholder handle. High-grading reached epidemic proportions as evidenced by miners coming off shift who had so many rocks hidden in their false pockets that if they stumbled getting out of the bucket or cage, they could not rise unaided. The toplander would often assist the miner in getting up but expected a piece of the action afterward.

Once the high-grade was on the surface, how was it sold? It could be delivered to one of several bars or to one of the assay offices in Victor or Cripple Creek. Major mining companies usually had their own assay office but a few independent ones were helpful to the very small miner or prospector. These legitimate assay offices would not melt high-grade ore for it would contaminate their equipment. However, there was a profusion of assay offices whose sole purpose was to buy high-grade from the miners, melt it down and ship it to a mint. If a bag of high-grade were left at one of the certain bars, a chit was

The El Paso headframe and sorting house. The latter provided one more opportunity for high-grading. *Photograph courtesy of Ed Hunter.*

given to the miner and about three days later he could return to the bar and receive his money.

Runners were hired by the bogus assay offices to make rounds of bars and rooming houses on a schedule to pick up the bags of high-grade. One miner reportedly had his wife wrap the high-grade in the clothes to be sent to the laundry. The laundryman then sold the high-grade and returned the proceeds with the clean laundry, less his cut for completing the transaction. However, the bogus assay offices paid for only half the gold content. One assay office cut the return to 25 percent. A few days later the owner was beaten up in one of the alleys and told to leave the district. This advice he gracefully followed.

Mine owners realized that a great deal of gold was escaping from their control. Estimates of the value varied greatly, but may have been well over $20 million at today's price. Detectives were hired by the owners to check on certain men suspected of high-grading and tracing their trips to the bars, assay offices or other drop points. Although the detectives wrote comprehensive reports, naming names, bars, assay offices and dates, very few cases were ever prosecuted, for it was not only difficult, but nearly impossible to obtain a jury that did not have one or more high-graders on the panel. Area merchants, who might be impaneled on a jury, were not adverse to high-grading, for it greatly increased the buying power of the miners leading directly to the merchant's pocketbook.

One judge even dismissed a charge of high-grading by ruling that a rock was real estate and therefore could not be stolen. In 1902, eight of the notorious assay offices were dynamited in the middle of the night, wrecking the offices and equipment. No convictions followed.

One mine owner ordered that metal lunch buckets were to be inspected at the end of each shift. Several miners then brought lunch in paper bags. The owner fired these employees but all employees then showed up with paper bags in place of metal lunch buckets. The owner recanted.

There appeared to be light at the end of the tunnel for the mine owners. Change houses were erected on the property adjacent to the shaft. All miners coming off shift had to go into the change room, remove their diggers, pass their lunch

The ubiquitous miners' lunch bucket used to bring high-grade to the surface until bucket inspections at the shaft collar became mandatory. *Photograph courtesy of Ed Hunter.*

buckets to a guard for checking, and walk without clothes to a second room where their street clothes were located. As the men moved to the second room, usually up a flight of steps, they were observed by company guards. Miners could be stopped and searched to make sure that no chunks of high-grade were secreted under arms or between legs or in other body cavities.

This was demeaning to the miners and work stoppages ensued. As a result, some mines let the men move from one room to the other in their underwear. Change rooms cut into the high-grading practice considerably, and were a controversial and bitter issue for the miners. Eventually, they learned to appreciate them, for they no longer had to walk home in wet clothes in subfreezing weather.

Underground miners are very resourceful and soon devised other methods for getting high-grade to the surface. Some men like electricians and timbermen could be persuaded to carry high-grade out of the mine with their tools and supplies that were not subject to searches. Another method was to place a bag of high-grade on top of a car of waste rock being sent to the surface for disposal. The topman who dumped the car would make certain that the bag ended up in a safe place on the dump so the miner could retrieve it after his shift. Old shafts provided high-graders access to active parts of the mine after the shift had been hoisted. Usually, the departing miners had blasted going off shift, so the high-graders would rush to the fresh muck pile and "pick the eyes" out of the pile.

Climbing in and out of the mine by old, rotten ladders was not all child's play and some high-graders even lost their way around the new parts of the mine. Harry Orchard, charged with the killing of two Vindicator Mine bosses and thirteen miners in the station explosion at Independence, was a self-acknowledged high-grader with experience in accessing old mines. However, he confused levels of the mine and set the Vindicator bomb on the wrong level killing the two supervisors rather than his intended target, the non-striking miners working on the next lower level.

The underground miner was not the only high-grader. The job of the ore sorters, stationed at a picking belt on the surface, was to pick any gold ore off the belt and drop it down a chute to the company storage bin. Some high-grade specimens seemed to migrate into the sorter's lunch bucket (which was not usually subject to search) rather than the company storage bin. Some sorting houses had peepholes in the walls for the ore house supervisor to watch for wrongdoings. However, some of these supervisors worked hand-in-glove with the high-graders.

The advent of the change room and the introduction of rustling or approved employment cards, which gave the operator the opportunity to void the card of a suspected high-grader so he was no longer employable, helped cut down on high-grading. In addition, legislation passed in later years required assayers to be bonded and to identify the mine from which any processed gold came. As the district mines went deeper, the veins became narrower and contained lesser gold values, which in itself reduced the opportunity for high-grade ore. This is not to say that high-grading did not persist to a more limited extent for many years.

Rufus L. Porter, miner, poet and historian of the district remarked, "high-grading violated the eighth commandment but a lot of the commandments had tough sledding in the Cripple Creek/Victor Mining District."

John Harper, Deported Miner: The Cripple Creek Strike as Family History

Katherine Scott Sturdevant

First, I need to make clear that the subject of my talk, John Harper, a deported Cripple Creek labor leader, was my great-grandfather, and that one of my specialties as a historian, as a history teacher, and as an author has been teaching genealogists how to do their family histories in historical context. So, this particular presentation is oriented toward the family experience in the Cripple Creek strike, and toward John Harper as a family man, as well as one of the mining leaders.

One of the practices essential to labor history is oral history interviewing. I was too late to know John Harper, let alone interview him. But we have oral tradition and family documents. With John Harper's experiences, one of the revealing exercises is to take what my family handed down as his experiences during the Cripple Creek strike, and see whether historical research confirms, expands, or refutes oral and family traditions.

The family oral tradition and documentation on Harper's experiences in the Cripple Creek strike is scanty. I think it became a skeleton in his family's closet. What happened at Cripple Creek in 1903 and 1904, as historian Elizabeth Jameson portrayed, could be a tragedy for a union miner and his family. That Harper was deported and blacklisted, did not take out of him the union-organizing spirit, certainly, but probably left the family doubtful about how much to share those experiences from then on with anyone else. Do you embarrass yourself if you share such a past? Do you perhaps even jeopardize yourself and your ability to be employed in the future by sharing it? Perhaps the Harper family feared so. Thus the family handed down the stories in bits and pieces, with pieces missing. It is interesting to see how the stories mesh with historical accounts.

So that you, too, might realize the great-granddaughter's experience, learning about John Harper for yourselves, let's start with what I learned earliest in my life about him. First came my mother's childhood recollections that she told to me. My

This picture of John Harper, 1904, was in the "mug book" found in Special Collections, PPLD, never before seen by the Harper family. We have not yet identified the apparently political buttons on his lapel. *Photograph from Special Collections, Pikes Peak Library District.*

mother was John Harper's granddaughter, through her father, Archie Harper, John's oldest son. When she met "Papa John" Harper—he lived in Arizona and she lived in California—he was an older man, a stranger, and she was a shy little girl. She said he had a very large nose, a very large moustache that hung down at the sides, a very thick Scottish burr that made him impossible for her to understand, and that he was very gruff and scary. When she had to sit on his lap he frightened her.

I remember thinking it must have been for her what it was like for me when she made me sit on Santa Claus's lap at the department store every year. I always cried. My mother's happier recollection about her grandfather John Harper at that time was that in some earlier life he had been a confectioner. He kept imbedded in his ceiling a large hook around which he still might throw salt-water taffy. No doubt this saved him, if nothing else would, from total ignominy with small grandchildren.

My mother's family stories interpreted John Harper as stern and moralistic. He would not have liked his picture here today being surrounded by pictures of the loose ladies from Jan MacKell's book. He also was a Presbyterian—a very stern, strict Presbyterian—and later in life an evangelical Baptist. In my mother's opinion, he tended to inflict those ideas very harshly on his family members.

These stories and opinions undoubtedly came to my mother from her mother, Kate, Archie Harper's wife, and thus are gospel according to how my mother raised me. There are several stories of John's disapproval toward his most "liberated" daughter and his more rebellious younger sons. When John Harper believed his grown children had sinned, he might insist they should kneel and beg his forgiveness along with the Lord's. When a misfortune befell a once-sinning child, John Harper might see divine retribution. Thus, I heard stories that did not endear John Harper to me.

Then there were stories of John Harper from my grandmother Kate's unpublished memoirs. She wrote about her experiences—from her 1880s–90s youth on a Kansas homestead, through marriage to Archibald Harper in 1904 Colorado, through their early days together with his parents in Arizona, circa 1906–1914. My grandmother married John Harper's eldest and prized son. After John was deported and blacklisted from Cripple Creek and Victor, he went to Arizona for copper mining. The Harpers lived in a place called Gibson's Camp near Bisbee. Kate, Archie's new bride, not entirely happy with their desert adventure or her in-laws, described John Harper at Gibson's Camp. She wrote about how the tent arrived:

> With three Harper men to boss the job of setting it up, you may well know there were interminable arguments. I'll always remember the day they almost came to blows about how the folding bed should be made. The tent top wasn't yet on the frame, and I got so disgusted with them that I walked far away. As I rounded the hill to pass out of sight in the desert, I looked back at a singing voice and saw John Harper, Jr. [the middle son], straddling the ridge pole, with his long legs dangling, singing at the top of his lungpower "Everybody Works But Father," [a popular minstrel song, circa 1905, about a father who avoided work at all costs]. Papa John stood below sputtering and snorting. When Papa was disgusted or angry, he had the funniest old snort I ever heard. He usually was snorting when his son John was around.

So, I had an idea of John Harper from family accounts. When teaching the use of family documents to genealogists, I used samples from John Harper's papers. My mother acquired these papers when her father, Archie Harper, died in 1965. They had been in Archie's possession since after his father John Harper's death in 1934.

John Harper's account book is revealing. In 1922 he made some money working in the "OD," which was the Old Dominion (Copper) Mine in Arizona. Then he spent some of that money on a list of books by Ian MacLaren, a Scottish minister who wrote sentimental accounts of village life titled after lines from Robert Burns's poetry. For 1926, the account book reveals some of John Harper's charitable donations: Presbyterian Church, Salvation Army, Moody Bible Institute, Williamsburg Jewish Missions, Stoneybrook Boys School, American Bible Society, New York Bible Society, First Christian Church, and Law Enforcement League. He often donated to Williamsburg Jewish Missions, because it was one of his missionary goals to "convert" the Jews.

Also among John Harper's papers are certificates that show how he invested in mineral stock as an old man: Texas Petroleum Company, Cordova Copper Company, Magna Chief Consolidated Copper Company, and so on. According to family tradition, John came to the Leadville gold rush in 1879 hoping to strike it rich. Do you think maybe, in the heart of the Great Depression, he still thought it was possible to get rich quick in mineral wealth?

All of these certificates, although they are all lovely and frameable, came in an envelope marked "other worthless paper" when he died. Oh, we did check. We have enough of him in us that we checked to make sure whether they had any value. So, even though, in the 1920s and 1930s, he was an older man who was no longer directly involved in mining, he was someone who was going to keep a financial interest in it.

One of my mother's cousins who grew up in Arizona with John Harper as an old man sent me some of his things in the 1970s, because she was becoming an older lady and I was becoming a family historian. I advise all would-be genealogists and family historians to do this: become the "keeper of everything," so that people see you as a repository when they are trying to dispose of their family things. Then prepare to live in an archival warehouse.

Cousin Kathleen (Grace Harper's daughter), who was my mother's generation, sent me odds and ends that had been John Harper's. One included a largely broken-and-glued-back-together teapot from Toronto, Ontario, Canada, that had imprinted on its side, "Droon yer sorrows in a cup o' tea." With this came a silver-plated sugar spoon he used, a framed picture of a sailing boat at sea, and the Scotsman's compensatory volume of Robert Burns's poetry. These were more clues to John Harper.

Now, we do have to ask, "Who was John Harper in a biographical sense?" Family history and historical research help. He was a man born in the Shetland Islands of Scotland in 1857. His father died when John was a baby. The Harpers and their neighbors were all fishing families who took to the North Sea. He came from a long line of John Harpers. It can make genealogy difficult—John Harper, son of John Harper, son of John Harper, son of John Harper, and so many of these John Harpers were lost in storms at sea that causes of death do not distinguish them.

Imagine what a tough life it was to be a Shetland fisherman or his wife. Many of the women lost their husbands, raised children by themselves, or remarried. They were a culture that believed in witches and selkies (mermaid sirens), and who counted upon different spirits to tell them their fate.

In *this* John Harper's case, his father had died before he could ever know him. His mother, Barbara Leslie Harper, remarried, to Samuel Kerr, a brick mason, had a second family, and they all moved to Canada. When he was in Canada, John worked at a confectioner's factory, where he met the girl who would become his wife. Her name was Katherine Margam. She was from Wales, and had come with her brothers to Canada. Later, some of the Harpers, Kerrs, and Margams went to Colorado and next, Katherine's brothers went to Pennsylvania, because they sought work in coal mining and tin milling. Ultimately, Margams tended to return to Wales, and Kerrs to Canada. John Harper, however, sought his fortune in Colorado.

In 1877, John (age 20) married Katherine (age 17) in Montreal. They had their first child named, guess what?—Herbert *John* Harper—in Canada, in 1879. That year, the Leadville gold rush was on. John Harper took his bride and their new baby in an open sleigh (some accounts say stagecoach) from Winnipeg, Manitoba, Canada, to Leadville, Colorado, in the dead of winter, to be part of that gold rush. After they arrived, the baby died of pneumonia in 1880. He lies in an unmarked grave in Leadville. So John Harper was apparently a tough man, a struggling man, a driven man, a man who might make some harsh decisions, and a man who might hope to get rich or at least find better employment in a gold rush.

Living in Leadville, John and Katherine had more children, starting in 1881 with the oldest son, Archibald Leslie Harper, my grandfather. By 1894, they had the five children they would raise to adulthood and had moved from Leadville to Victor, following the rushes. John worked for mines—in Victor, he was in the Gold Coin Mine. He was successful enough that they lived rather well. He raised his children to value education. Archie graduated Victor High School, was class valedictorian, and entered the University of Colorado Law School in Boulder. Later, middle son John Herbert Harper would attend the Colorado School of Mines in Golden and become a mining engineer. In Victor city directories, it's apparent that even the girls (Grace and Pearl) were learning careers, so that one was going into secretarial work—business school—and the other was going to become a teacher. Katherine seemed to have some finery.

In Betsy Jameson's book, *All That Glitters*, is a chapter discussion of women's roles in the mining community. It adds context to the life of Katherine Margam Harper. Family tradition tells us that she was a quiet, sweet, gentle lady. She was Welsh, and many characteristics often associated with the Welsh were true of her. She loved to sing. She loved to play piano and, by this time in Victor, she had her own instrument—a lovely old upright piano—to play. She stayed home. She seemed to be the epitome of the immigrant's wife in the domestic sphere. She didn't complain, she didn't criticize, and she got pregnant every couple of years. Family women's oral tradition is that she was getting pregnant too often for her own good, and that the family doctor spoke to her, and apparently to John, about this.

Stories of private life sometimes survive woman to woman. I am the only daughter of my mother, who was the only daughter of Kate, who married Katherine's son Archie. Katherine confided to her future daughter-in-law, my grandmother, before their wedding. The doctor said it was going to kill Katherine to have more babies and that her husband should leave her alone. He didn't, and she got pregnant again. So then she took her children on an extended trip to visit her family in Wales, for two years. Of course, she didn't get pregnant during that period. When she came back, she did immediately, but that would be the last known child that she had.

Katherine Margam Harper, ca. 1902, the respectable wife of the president of Victor Miners' Number 32. *Photograph courtesy of Katherine Scott Sturdevant.*

Still more personal was the story Katherine told Kate about her own feelings. She said she lay in bed every night and listened to her husband turn off the gas, come up the stairs, Katherine dreading with every footstep the "awful thing" he would do to her when he came to bed. It's the kind of story, again, we would not normally know—she certainly would not have written it down. Family women's tradition even gives it a punch line. When Katherine told that story to her future daughter-in-law, Kate, as a warning about men and what they do, my grandmother Kate went to *her* mother Fannie, who was a Midwestern, Scotch-Irish, homesteading woman, and told Fannie this story. Fannie announced to her daughter, "If he'd been my husband, I would have locked him in the barn!"

Publicly, John Harper was the kind of man that Betsy Jameson describes miners as being. He was "a pillar of the

Presbyterian Church," according to *Miners' Magazine* (the Western Federation of Miners publication). His wife was a domestic lady, and his children were stellar in school. He was very active in the WFM as an officer and committee man, as well as a political candidate. He became president of Victor Miners' Local No. 32 in 1902 and 1903.

What caused him to end up in the center of the dramatic events of 1903–1904, however, was not so much having been president of the local, but what he undertook next: he became the manager of the Victor Cooperative Store that the miners' union set up to support the striking miners and their families. When the strike was declared, the miners had thought that the community businessmen would keep their promise that striking miners could buy whatever they needed on credit. Of course, the businessmen not only failed to keep the promise, but they formed organizations of their own, the Mine Owners' Association and the Citizens' Alliance, aimed at union busting.

Cooperative stores were one of the main tactics of the Western Federation of Miners to keep the striking miners going. You could buy at cost, you could buy on credit, and you were not putting your money into the coffers of the businessmen opposing the strike. Most of the *Miners' Magazine* accounts cite John Harper specifically as one of the intelligent managers—my Scottish relatives might have said "canny wi' a penny," the old stereotype of Scotsmen being good with money. John Harper managed the Victor store while another man, T. H. Parfet, managed the Cripple Creek store. Both became so good in this role that they also became obvious targets for the businessmen, who would resent them to the point of violence.

John Harper was one of the WFM leaders quoted as saying "Is Colorado in America?" or "Is this a free America?" It was a common remark made in Western Federation of Miners rhetoric during the strike. It's very clear that the interest in harassing and arresting these men had to do with their leadership in the miners' locals and not necessarily with any accusation of violent activity. One of the oral traditions handed down in my family was that John Harper disavowed violence. I don't know whether this was a source of pride for John's immediate family, or whether the reason my mother kept telling it to me was because my mother had become a Quaker, and so his pacifism would appeal to her. According to family tradition, during the strike, a minister came armed to the Harper home, and told John Harper, "You need to have this gun. The men are going to come for you. The night riders will come for you and so you need to protect yourself and your family." John refused the gun and said he would never resort to violence.

At this juncture, perhaps I should explain myself, although some of you know this about me. I decided to mention it partly because a member of this audience approached me about it here in the library lobby. One reason I have "UNION" emblazoned across my chest (with a rhinestone pin) is because I am the president of a small local of a teachers' union. As such, I experienced a modern version of retaliation. Unlike John Harper, I was not arrested, jailed, kidnapped, physically attacked, threatened with lynching, deported from the state, or blacklisted from future employment. The retaliations, however, seemed no less dramatic to me, in my modern workaday world. Many colleagues and acquaintances wondered that I did not quit my job and flee Colorado altogether. Also unlike John Harper, I was successful in my case, but only after 5–6 years of distress. Ultimately I felt fully vindicated. Knowing this about me adds context and meaning to following my discoveries about what John Harper went through.

As my husband Rick could tell you, when I was sitting in the University of Colorado, Boulder, archive reading the following, I had a reaction that caught both of us off guard: "John Harper, former president of Victor Miners' Union grocery store in this city, today was released by order of the military from confinement in the city jail, where he has been held for a little more than three weeks. His release was granted in order to give him an opportunity to leave the city or be forcibly deported. He chose the latter alternative. He refused to surrender his membership in the Western Federation of Miners, as dozens of others have already done, and this is the prime cause for his deportation." When I read that—and it happened to be the first really substantial paragraph in the records that I had found about him—I started to sob. The archivist did not quite know what to make of a researcher crying over an old *Miners'*

Magazine. My parallel connection with John Harper taking his stand had struck me across a century.

John Harper became one of 238 men deported. We are not sure yet whether he went to the Kansas prairie or the New Mexico line, but he was ordered not to return to Colorado on pain of hanging, because he would not give up his union card.

Of course he was also one of the men held in the bullpen. At one point, at least, he was "sweated." The first time I ran across that news, "We sweated John Harper," I thought it must be a euphemism for interrogating someone. I had no idea how literal and nightmarish it was. They used sweat boxes, into which men were shoved before questioning them—torture to see if they could get men such as Harper to tell more. In his instance, they wanted him to reveal whether there were guns stashed in the store, where the guns were, and who might have used them.

When Victor and Cripple Creek were under martial law, after the union was broken, John Harper was somewhat safe because of the martial law. When the governor pulled out all of the troops, the problem became those resentful businessmen, who could take the law into their own hands. The cooperative store raided, destroyed, cans turned upside down, the staff arrested by an illegal mob—western vigilantism overtook the town.

Among the August 1904 retaliations that John Harper suffered is one series of events that relates to a family oral tradition. The family recalled an occasion when men came to the Harper house and took John away. By the time I heard the story, the family did not describe the men or know who they were. They said that this was one of several times that John was taken away—that he was beaten up and that he had to find his own way back home. Interestingly, they never described to me the emotional family members' perspective on this experience, which *Miners' Magazine* tells very graphically. Perhaps it was so painful or embarrassing or nightmarish that the original tellers in the family censored the story.

Harper and Parfet, the two managers of the Victor and Cripple Creek cooperative stores, set out to return and close up the stores. The sheriff had promised them protection from anyone—such as a secret nighttime mob coming from the

Citizens' Alliance—but, of course, that protection would not be forthcoming. John Harper was at home.

> At eight-thirty o'clock, he was talking with his wife and his son Archie, age twenty-one years [Archie didn't live at home at the time; he was a student at the University of Colorado Law School at Boulder but was apparently visiting], and two daughters, Grace and Pearl. Mrs. Leland, wife of the Methodist minister, was calling on the family at the time. Mrs. Leland says the men grabbed Harper and his son. Someone asked, "Where is that rope?" [These "whitecaps" were hooded and masked in the tradition of night riding that we associate with the Ku Klux Klan.] One of the masked men replied he had forgotten the rope but would go get one. Harper and his son were hurried outside, and Mrs. Harper ran after her husband, threw her arms around him, and kissed him. One of the men roughly pushed her to one side. She wrung her hands and asked to be permitted to give him a last kiss before he was strung up. Mrs. Leland says that one of the men struck Mrs. Harper and remarked, "No more kissing now!" After Harper was taken away there was a panic in the home. Archie was permitted to return, but the girls and Mrs. Harper are prostrated and require medical attention constantly."

Again we see a vignette of Victorian womanhood. What happened to the Harper men that night? The whitecaps dragged John and his son Archie down toward Cañon City. Before doing any other harm to John, they decided to release the son. Two of the men escorted Archie back. He was able to return home and say no harm had come to his father. After they released Archie, however, they beat up John. They left him in Cañon City.

> John Harper, who was beaten by a mob and driven from his home last night, will be in Denver this evening and report to Secretary Haywood. He called his family and told them he was alright. Concerning his deportation Secretary Haywood said, "Harper had permission to return to the district to close up the shop, and he

> had a guarantee from Sherman Bell that he would be protected if he did come back. After the mob had beaten and chased Harper from the district, he waited in Cañon City, and he will take the train to Denver." Haywood had a conversation over the telephone with Harper this evening and said he was beaten about the head and face and robbed of the money that he had with him, which was ten or twelve dollars.

So research clarifies family tradition. Research also reveals Harper's importance in the strike—that he telephoned "Big Bill" Haywood, who obviously knew and respected Harper.

What impact did all of this have on young Archie Harper, my grandfather? In *The Silver and Gold,* the little magazine that the University of Colorado at Boulder printed for its alumni, edited by Archie's law-school classmates, we found this note: "Arch Harper of present senior class was deported from Cripple Creek this summer and thus far has been unable to get back within speaking distance of the university, but will be back next year when the Western Federation of Miners has hopefully gone out of existence." Apparently his friends at college just could not understand why he was doing these things.

Archibald L. Harper graduated University of Colorado at Boulder Law School in 1904. His classmates' favorite quote about him was from Robert Burns: "A Man's A Man for A' That." *Photograph courtesy of Katherine Scott Sturdevant.*

My mother's oral tradition about her father Archie's experiences is interesting. She said that Archie was a student at the law school, that he came home to visit his parents on the train, that he was arrested when he got off the train, and that the reason he was arrested was bogus. The reason supposedly was that, when he was back in law school, they had conducted a class debate over the issue of the strike. Supposedly Archie was assigned by the professor to take the miners' side in this debate, and when the people back in Victor found out that he took the miners' side in a debate at law school, he was arrested.

What do you think? Could this version be true, or was the family finding ways to remove Archie from any culpability in his father's union activism? Well, family tradition and records verify that Archie worked in the mines to put himself through law school. Records further note that he was a member of Victor Miners' Local No. 32 and served on a union committee in 1902. His classmates addressed his WFM activities as though they were foreign and distasteful to them. So Archie was more directly a union man than family tradition allowed.

John Harper was deported finally and blacklisted from future employment in Colorado mining. These photographs of miners that we have displayed here for the symposium were apparently "mug shots" of deportees intended to help blacklist them. It was so exciting when Chris Nicholl informed me that the library had this collection and that she found John Harper in it. No one in the family had ever seen this picture of him—nor any picture so revealing of his facial character. Comparing this photograph with one we had from just two years before also portrays how much stress aged the man. Yet he has a proud look, doesn't he?

When John was blacklisted, the Harpers had to leave the state. Family tradition tells of how painful it was to leave their Victor home. The house was a symbol of how much they had achieved, yet they had to sell it for just $125. It broke their hearts, they said.

The one thing Katherine would not leave behind—the piano. Thus there are many family stories about moving the piano, loading it into a wagon and taking it down themselves to Globe, Arizona. They had to abandon nearly everything else, making small mementos (such as family photographs or a teapot from Canada) very special. They went to an ignominious place, Gibson's Camp in Arizona (copper mining near Bisbee), where they had to worry about scorpions and child-of-the-earth bites.

Young Archie suffered lung damage from working in the copper mines (and later from smoking), so that not only could he never work in mining again, but he had trouble ever visiting

Colorado again because of the altitude. Leaving Colorado in 1906 started a series of hardships and misfortunes for Archie that ruined what had seemed a bright future. When he had first become an attorney in Colorado, he worked with Judge Ben Lindsey, the nationally recognized Progressive who reformed the legal treatment of children and was the father of the juvenile court system. Lindsey sent Archie to represent the firm at the trial of Big Bill Haywood and the other WFM leaders in Boise.

Later in life, Archie told what it was like to watch Clarence Darrow defending the miners. After blacklisting drove his parents out of Colorado, however, Archie at first investigated Goldfield, Nevada, as a possible home, but found the same industrial warfare there that had just occurred in Victor. So he went with the Harpers to Arizona with his bride Kate. He opened a law practice in Globe, specializing in mining law, and ran for district attorney.

According to family tradition, when Archie campaigned across frontier Arizona on horseback, however, the local press blackened his name as a Colorado labor agitator. He lost the election and his reputation. The political party "hacks," including the man who had been elected governor, wanted to give him some kind of consolation prize.

So, the governor appointed Archie the head of what was called the Arizona Territorial Industrial School—a reform school heavily populated with Mexican youth. Soon, he faced scandals there, because when the students rioted, he used corporal punishment. The family tradition was that he used a strap on the ringleaders who had threatened the lives of his wife Kate and little son, John Leonard Harper. The local press accused Archie of cruelty.

Archie Harper suffered a "nervous breakdown" as he faced the impending trial that never happened. Correspondence he kept with his attorney reveals what an ordeal this was.

On doctor's advice, Archie and family took a rest in California and, after they returned to Arizona, moved to Canada for a lesser job and home with the assistance of Kerr relatives. He was in charge of accounts for International Harvester and thus collected bills. Proud stories include how he outfoxed cagey debtors, with references to his Scottish heritage again, of course.

Kate had lost one baby, Marian Katherine, in 1908, then had John Leonard ("Len") in 1909, so Len's earliest years were in Arizona. But her last two were born in Saskatoon, Saskatchewan, Canada, a place she found most remote and lonely. Leslie Archibald was born in 1915 and Barbara Pearl (my mother) in 1917. Soon the family moved to Vallejo, California, near Kate's relatives, where Archie spent the rest of his life as a bureaucrat, though a leader in Masonic Lodge and Presbyterian Church. So Archie's children grew up knowing that their father had been an attorney, without understanding why he quit, but in awe of his integrity and bearing.

Archibald Harper and his granddaughter Katherine Scott (Sturdevant), in 1955. He lived until I was 10 and was a much revered "Papa." *Photograph courtesy of Katherine Scott Sturdevant.*

Here is a picture of an older man with a baby. That's Archie Harper and me. Archie Harper was my "Papa," my grandfather. He lived until I was ten. He certainly never told me stories about his father; my mother did that. She named me Katherine after his Welsh mother, to please him, and he called me "Katie," after his wife. I hero worshipped him, sitting on his lap or at his feet. He commanded respect, although there was an unspoken sense of failure or regret. In the face of challenges to his character and integrity, he gave up the law and fled the fight. After all, his father John Harper had stood his ground, yet lost his fight in spite of his courage, and fled Colorado anyway.

My mother, perhaps under the inspiration of her father and my love for him, raised me to think about him, his values, and his father's values. As I grew up in San Francisco, there were copies of Burns's poetry and MacLaren's books all over the house (my father was Scottish too). My mother talked about what Colorado meant to Papa, how he longed to return, making it sound beautiful. John Denver's songs helped. Rick and I planned to live in Colorado some day.

In 1985, Rick got transferred from the Midwest to Colorado Springs. I thought, rather casually, that Colorado would be nice: I've got some family history there. That would be interesting to investigate someday. I really had little clue. Remember, I had heard snippets of oral tradition and had felt nostalgic impressions, not the whole story of the Cripple Creek strike and John Harper's role in it, or how deeply it affected the family.

I came to Colorado Springs and started to teach. With my San Francisco and union-family background, I naturally joined the little local teacher's union. Then I naturally became a little local leader of it. Then I naturally got in a little local trouble. It did not take too much, because this is Colorado and "*is* Colorado in free America?" I was naïve about Colorado's continuing resistance to labor activism. I found myself restricted and retaliated against for what seemed mild exercises of constitutional freedoms.

Of course, I responded in the modern, civilized, pacifist way—no dynamite, no guns—championed by wonderful lawyers (I've always loved lawyers for some reason) through Colorado's federal court system. But, at first, I only vaguely thought about a connection to John or Archie Harper. The strongest relationship I recognized soonest between my mother's family history and what I was going through was this: I could never quit and flee like my beloved Papa had done. I had to prove I could withstand the circumstances and stay in my chosen place—professionally and geographically.

As I suffered punishments and character assassination, I serendipitously discovered a bit more about John Harper's Colorado experiences. First, Rick found one day, in the Colorado College Tutt Library, that Emma Langdon's book, *The Cripple Creek Strike*, mentioned my great-grandfather by name and described the events he experienced. We sought and splurged on an original copy of my own to treasure.

Then I'll never forget standing with Rick in Barnes and Noble bookstore when, on a second hunch, we picked up Betsy Jameson's book and found John Harper mentioned, quoted, and indexed. As I read and researched, a connection with my great grandfather became apparent. I realized that I could take some solace in family heritage. As I read more about what he went through, I decided there was something admirable, even heroic in the man. It dawned on me, too, that my mother's family had been driven from Colorado, but, ironically, I had returned and stayed.

Sometimes you find your moment in history—that moment when you must take a stand, a stand that might even be of historical significance, whether or not anyone else knows you're doing it. It's a stand that shows what you're made of, that tests your mettle. It's a stand that requires you to have courage, to see how deep your ethics and integrity go. It's a stand that tests your willingness to sacrifice, that demands the sacrifices of your family on your behalf. It's a stand that involves risk, and whether the risk is being deported, or lynched, or jailed, or losing a job, you take it. The courage to take a stand, whether it comes from Presbyterian training or Quaker faith, from growing up in the Shetland Islands or San Francisco, or from loving a Papa who had given up his stand, it stiffens your spine when faced with wrong. It connects across a century.

Thank you for honoring people of the past who had the courage of their convictions, and kept their integrity, even though they failed. The miners did fail to achieve their goals in 1903 and 1904. I succeeded, but the success I learned to savor most is the success that I think John Harper also achieved and carried with him the rest of his life—he took a stand, and in the face of all threats he kept his conscience.

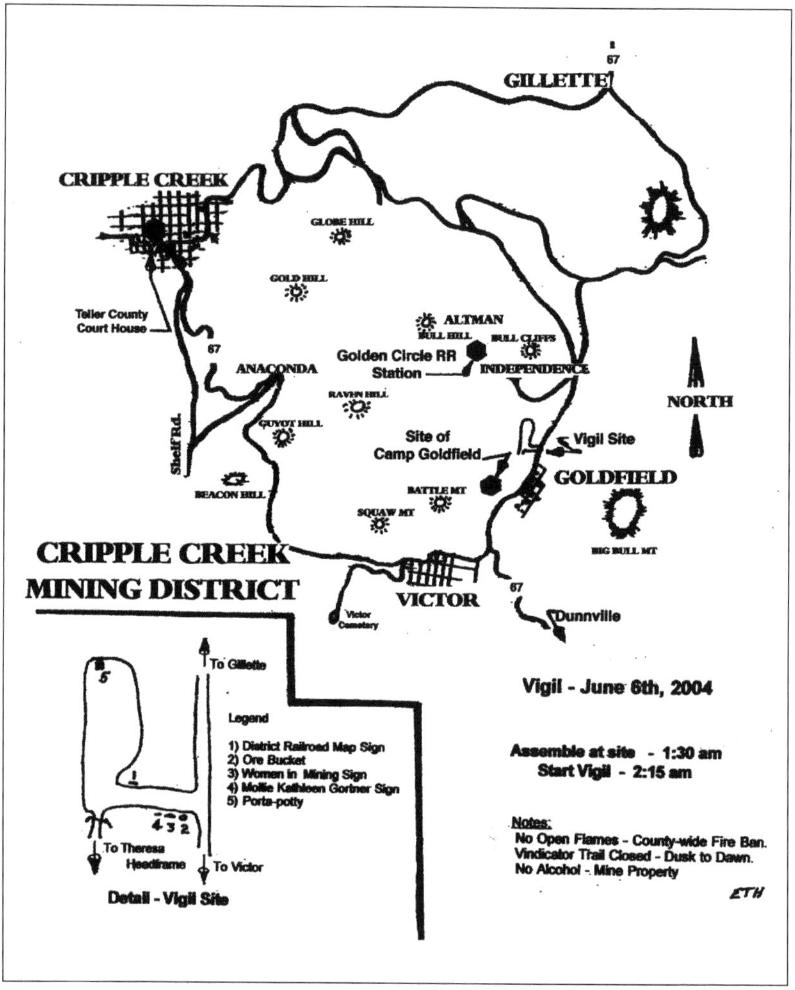

Map 1, Cripple Creek Mining District, vigil site, 1904 Strike Commemoration. *Courtesy of Ed Hunter.*

Talking the Walk: The 1903–1904 Strike Centennial in the Cripple Creek District

Elizabeth Jameson

The tragic climax of the 1903–1904 strike came on June 6, 1904, with the Independence Depot explosion, the forcible takeover of local government by the Citizens' Alliance (CA) and the Mine Owners' Association (MOA), the Victor riot, and the destruction of the local union halls and cooperative stores. On June 6, 2004, the centennial commemoration moved to the original sites of conflict, with a vigil near the site of the devastating explosion and a walking tour of the strike scene in Victor the following afternoon.

The Vigil

At 2 A.M. June 6, 2004, a small group of stalwarts gathered as close as possible to the former Independence townsite, where, a century before, twenty-seven non-union miners who had just come off shift waited on the platform of the Florence and Cripple Creek (F&CC) depot for the 2:15 A.M. train (see Map 1). Two worked on the Deadwood Mine, the rest at A. E. Carlton's Findley. Just before the train arrived, an explosion ripped through the depot platform, killing thirteen of the men and wounding six others. The explosion that killed them also began the end of the strike, and of the district's vibrant union community.

We who gathered to remember could not begin to imagine the scenes a hundred years before, the town that had been there, the mutilated bodies, the impassioned conflict that led to those final acts of the bitterly contested strike. We could only bear silent witness to the men who died that night, and to the others on both sides whose lives were forever marred.

Mindful of the high fire danger that dry chilly evening, we lit thirteen electric candles for those who died at the Independence Depot, to reconnect in memory the men the explosion literally dismembered, to remember the men whose tragic deaths have been too easily reduced to a marginal note in history. Their names have been all but forgotten; some were never listed in the report of the coroner's jury that held an "inquisition" "upon the bodies of Gus Augustine, Arthur Muhliesen, Henry Haag, Herbert McCoy, William Shanklin, Edward Ross and others lying there dead."[1] We don't know what brought them to work during the strike in a non-union mine, or to be standing in exactly the wrong place at exactly the wrong time. We could only briefly interrupt our sleep to remember thirteen men who, for unknown reasons, worked the night shift on non-union mines, and who died, pawns in a conflict waged to win other working men's rights.

Vigil at 2 A.M. on June 6, 2004, near Goldfield, Colorado. Left to right, Rick Sturdevant, Kathy Sturdevant, Ed Hunter, Elizabeth "Betsy" Jameson, Bridget Burke, Chris Nicholl, Nick Nicholl, Marcia Goldstein, Jeff Goldstein. *Photograph courtesy of Bruce F. Davis.*

The Walking Tour

The following afternoon a larger and more rested group toured some of the major strike sites in Victor (see Map 2). Here, I will describe the sites in the order that we visited them, which follows the geography of the walking tour rather than the chronology of events.

Baltimore Hotel Site, (map location 10), 103 North Third Street. The night of June 5, 1904, most of the militia and prominent members of the Citizens' Alliance stayed at the Baltimore Hotel in Victor. A. C. Cole, who had been a Victor High School teacher and a prominent district Republican and who had served as secretary of the Victor Citizens' Alliance from December 1903 to June 1, 1904, later reported what happened at the Baltimore Hotel that night. The Citizens' Alliance was mobilized as Company L of the Colorado state militia. Cole served as Second Lieutenant of Company L during the strike, and was present at the Baltimore Hotel in that capacity. Few slept, he said, "in anticipation of some event." Most were awake when news of the explosion came. They reportedly said that Edward Bell would be made sheriff when the trouble happened. The militia was instructed to clean out the Western Federation of Miners (WFM) members when "the shooting began" the afternoon of June 6. "It was generally understood and freely discussed that a riot was to be precipitated," Cole charged.[2]

Site of Open Field, 1904, (map location 6), North Fourth Street and Victor Avenue. Over the objections of the county commissioners, the MOA and the CA held a public meeting in a vacant lot across the street from the Victor Miners' Union Hall the afternoon of June 6. When the meeting was called, the militia, MOA, and CA had already halted Sheriff Henry Robertson's investigation of the explosion, and had seized the victims' bodies from the elected coroner, M. J. Doran, a member of organized labor, insisting that he surrender them to J. H. Hunt, a local undertaker. The site of the meeting, and the speed with which they had moved to depose local officials suggested the likelihood that the meeting would provoke violence. Before the meeting was held, Sheriff Robertson was forced to resign and was replaced by Edward Bell.

Victor City Marshall Michael O'Connell, hoping to stop the meeting, swore in some hundred deputies. But Mayor F. D. French, one of the few local officials not aligned with organized labor, fired O'Connell, revoked his deputies' commissions, and replaced him with Maj. H. A. Naylor, temporary commander of the militia.

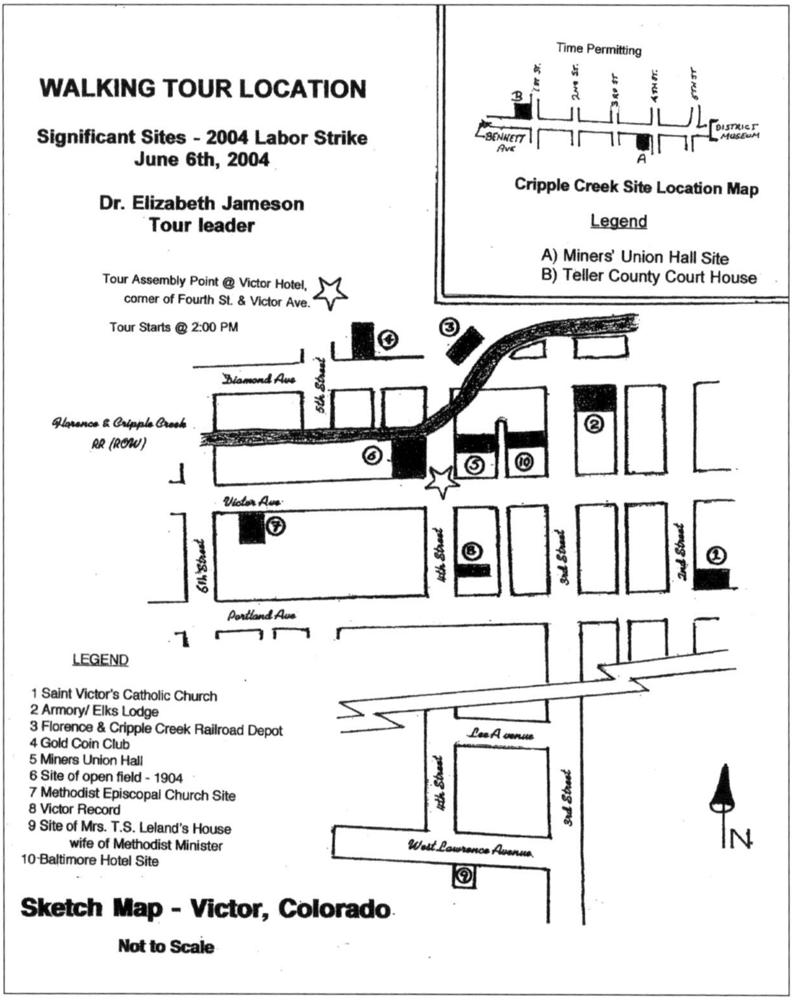

Map 2, Victor walking tour. *Courtesy of Ed Hunter.*

A crowd of several thousand gathered around the vacant lot and union hall. MOA secretary Clarence Hamlin addressed the crowd, blaming the WFM for the Independence Depot explosion and demanding that the union be destroyed. "The badge of the Western Federation of Miners is a badge of murder," he declared, "and everyone who is responsible for the outrage at Independence should be driven from the district."

Alfred Miller, a union miner, spoke up to challenge Hamlin. Miller was carrying a rifle. He was standing next to his brother Christopher, who had, incidentally, earlier been deputized by Sheriff Robertson and assigned to guard evidence at the explosion site and then been relieved of his responsibility along with Robertson. Fearing trouble, Chris Miller tried to take his brother's gun, but his action was misinterpreted as a hostile gesture. In the ensuing riot, two men, Roscoe McGee and John Davis, were killed and five more were shot. Around fifty union miners fled to their union hall across the street.[3]

Victor Miners' Union Hall, (map location 5), 110–112 North Fourth Street. The Victor Miners' Union Hall was the site of the first direct attack on organized labor on June 6, 1904. As soon as the first shots were fired on North Fourth Street, between the vacant lot and the Miners' Union Hall, Sheriff Bell called out Company L, commanded by Lt. W. W. Travell, manager of the Stratton's Independence Mine. Company L had assembled nearby at the Victor Armory, and arrived immediately. The militia fired on the hall, and shot four miners before the union men surrendered. The Citizens' Alliance then invaded the hall, wrecked it and destroyed union records, and sent more mobs to loot and wreck all the WFM halls in the district and the four WFM cooperative stores. By midnight, they had vandalized every union store and union hall in the district. Standing outside the hall, the walking tour participants could still see some of the bullet holes left from the battle a century before.

The attack on the Victor Miners' Union Hall was strategic. In fact, the union leaders later accused in the Independence Depot explosion were not from Victor WFM Local Union No. 32. Those accused (but never charged or tried) were officers of the national WFM and of the Altman WFM local, Free Coinage

No. 19. Victor Local No. 32 was, however, the largest local union affiliated with the Western Federation of Miners, and the largest local of any union in the state of Colorado. Founded in late 1893, shortly after the Altman local, Victor Miners' No. 32 operated under the charter of Altman No. 19 until after the 1894 strike was settled. It obtained its own WFM charter January 12, 1895. By 1902 it claimed over one thousand members. Even in 1904, after many men had left the district during the strike, its membership was over eight hundred.[4]

In the late 1890s the members of No. 32 began raising money to build their own hall. They bought the lots on North Fourth Street in October 1900 and dedicated the new hall in March 1901. The total cost of the building and land was $30,000; by 1902, Victor Miners' No. 32 held its building debt free.[5] It rented out two retail spaces downstairs. The upstairs held a large auditorium and the union offices. The windows through which both sides shot were in the upper front of the building, in the office space.

After the district WFM locals were destroyed, the national WFM owned the building. The WFM, which changed its name to the International Union of Mine, Mill and Smelter Workers in 1916, gave the union hall to the Victor School District in 1924. The upper floor was remodeled as a gymnasium, and a stage and locker rooms were installed. The main level housed a manufacturing facility during World War II, and later a movie theater and restaurant. The school moved out in 1969, and the building has had several private owners since then.

Owners Joe and Marjie Stevens kindly allowed the walking tour to go inside the building, which is in extreme need of repair and renovation. Not until we were inside, and could see how deep the building is, could we grasp how large and impressive a structure it had been. It clearly had belonged to a large and stable local union. The smokers, balls, political and union meetings held there would have been large affairs; the hall would have been a central site of working-class community. To attack the Victor Miners' Hall would have been to attack a major symbol and center of union power. And it was just around the corner from the Armory.

Gold Coin Club, (map location 4), Diamond Avenue, between North Fourth and Fifth Streets. The Gold Coin Club was adjacent to the Gold Coin Mine, owned by Warren Woods and his sons Frank and Harry. In 1893, the three organized the Woods Investment Company, which began selling town lots in Victor, below the rich Battle Mountain gold deposits. They struck accidental paydirt while digging the basement for a hotel in 1894; that lucky strike became the Gold Coin, the first holding in an economic empire that eventually included the Economic Gold Extraction Company, the Pike's Peak Power Company, the Golden Crescent Water and Light Company, the Teller County Mining Supply Company, the United Mines Transportation Company, twenty-four mining companies, and the Colorado Springs and Cripple Creek District Railroad, known as the Short Line.

Devout Baptists committed to the stewardship of wealth, the Woods believed that investing in their employees' social welfare would promote smoother industrial relations. In that spirit, they organized the Gold Coin Club in 1898 and erected the clubhouse opposite the Gold Coin Mine in Victor. Employees who worked six continuous months for the company could belong. The original Gold Coin Club House was completed only to be burned in the Victor fire of 1899, which destroyed much of the downtown. Reconstructed at a cost of $40,000, the Gold Coin Club was dedicated in February 1900 at an elaborate party with dancing and a banquet.

The two-story Club House housed a library, an athletic room, a bath room, where men could bathe when they came off shift, a billiard hall and chess tables that offered games at half the price charged at local saloons, and a few bedrooms for office employees. Jack Johnson worked as a porter at the Gold Coin Club, and a number of less talented fighters entertained crowds there.

The Woods' amicable relations with their employees ended during the 1903–1904 strike, when the Woods Investment Company caved to pressure from the MOA to fire union miners. On September 30, 1903, Frank Woods announced that all company employees must quit the union at the end of the day shift. They walked out instead, including the superintendent, shift bosses, and all workers at the Woods' Economic Mill.[6] After the strike, the Gold Coin Club building was, for a time, a hospital. It is a reminder of the respectful relations between mine owners and union labor before the strike.

Florence and Cripple Creek Railroad Depot, (map location 3), Diamond Avenue and North Third Street. The Victor depot of the Florence and Cripple Creek Railroad (F&CC) was on Diamond Avenue, near the Gold Coin Mine and just below many of the big producing mines on Battle Mountain. It was also across the street and a half block from the Colorado Trading and Transfer Company, owned by A. E. Carlton. Carlton had begun his business career in the district hauling ore and freight, and had then begun to expand his holdings to mines like the Findley. The Victor Armory was across the street from the Trading and Transfer Company, and diagonally across Diamond Avenue from the F&CC Depot.

The F&CC Depot was a significant strike site because of its strategic use by the employers and the state militia. On the morning of June 7, 1904, Clarence Hamlin announced the employers' plan to deport union miners. The MOA and CA appointed a seven-man commission to decide who could stay and who must leave and began interviewing the union men they had seized. Prominent leaders were slated for deportation. All others, to have a chance to stay, had to renounce union membership and accept MOA cards that would be required to work on district mines. At 5 P.M. deputy sheriffs took twenty-eight men—none of whom were charged with any crime—from the Armory to the Victor F&CC Depot, put them on a train, and deported them to Denver.[7]

When he arrived in the district, Adj. Gen. Sherman Bell approved the deportation procedures and converted the examining committee into a "military commission." By late July, the commission questioned 1,569 men, recommended that 238 be deported, that charges be filed against forty-two, and that 1,289 be released. The militia loaded the deportees onto trains, took some to Denver and others to the Kansas and New Mexico state lines, unloaded them, and ordered them to stay away from the district.

Immediately after the first trainload of deportees left the district, General Bell commandeered another train to launch an attack on some sixty-five union miners who were prospecting at Dunnville, a new mining camp eight miles south of Victor. On June 8, Bell, thirty soldiers, and a hundred deputies took an F&CC train to Dunnville to seize the prospectors. The miners had a total of two rifles, three shotguns, and five revolvers. The fully armed troops and deputies fired on them, and in the brief gun battle, John Carley, a union miner, was killed. Fourteen others were arrested and brought back to Victor on the F&CC train.[8]

Armory/Elks' Lodge, (map location 2), 181 North Third Avenue. For some years the home of the Victor Elks' Lodge, 181 North Third Avenue was, in 1904, the National Guard Armory. As such, it played a crucial role in the military occupation of the district, and in the events immediately following the Independence Depot explosion. Company L member and former CA secretary A. C. Cole reported that before June 6, 1904, the Victor militia held an "unusual number of drillings," that considerable ammunition was brought to the armory on June 5, that Capt. Harry Moore came to the armory that night to check that the arms and supplies were in instant readiness, and that the militia was ordered to prepare for instant call. After the explosion, General Sherman Bell activated local units of the National Guard, putting Maj. H. A. Naylor in temporary command. The local Citizens' Alliance National Guard unit, Company L, assembled at the armory. For weeks thereafter, the armory served as headquarters for the militia and the mine owners.

The morning of June 6, it was the location of the first forced resignations of elected civil officials. After the explosion, Sheriff Henry Robertson had rushed to the scene with all the deputies he could find. When he arrived at the Independence Depot, Robertson roped off and searched the area, questioned witnesses and nearby residents, telegraphed other counties for bloodhounds, and swore in some hundred deputies. His investigation uncovered about seventy-five yards of wire with one end wound around a chair leg and part of a revolver, presumedly parts of a device that exploded 150–200 pounds of dynamite.

News and rumors spread quickly. All the mines closed except the Portland, which was operating with union labor. The district split into opposing camps, those who assumed that the WFM was responsible, and those who presumed it was innocent. As armed and angry men assembled on the streets of Victor, Robertson ordered the saloons closed.

The Mine Owners' Association and Citizens' Alliance gathered at Victor's Military Club in the armory, determined to depose elected officials whom they considered pro-labor. They began with Sheriff Henry Robertson. When Robertson came voluntarily to meet with them, the employers demanded his resignation. He refused. They fired several shots, and offered him the choice of resigning or being hanged. Robertson resigned. The owners then persuaded the county commissioners to replace him with Edward Bell, a member of the MOA and CA, whose $10,000 bond they had already arranged. Sheriff Bell immediately fired Robertson's under sheriff, James Gaughan, and replaced him with Citizens' Alliance secretary L. F. Parsons. Within days, the MOA and CA forced over thirty officials to resign, imprisoned many of them, and replaced them with anti-union appointees.

Published versions of Robertson's "resignation" always mentioned "a coiled rope." A photograph at the Colorado Historical Society, however, shows a man with a noose actually around his neck surrounded by an armed mob. The room in which the photo was taken is clearly the hall on the second floor of the armory, still recognizable when we visited the Elks' Hall on our walking tour. Though I cannot be certain the man with the noose around his neck was Robertson, the scene matches accounts of the confrontation with Robertson at the armory (see photograph on page 11 this volume).

After Robertson's "resignation," members of Company L waited at the armory until Sheriff Edward Bell called them out after the first shots were fired outside the Victor Miners' Union Hall. After the battle between the militia and the union members inside the Hall, the union men surrendered to the militia and were taken to the armory. The armory served as one of the military "bullpens" in which union members were held before being questioned by the "military commission."[9]

Lowell Thomas Museum, (not marked on the map), Victor Avenue and North Third Avenue. The walking tour stopped briefly at the Museum, named for Victor native son Lowell Thomas. The Museum has interesting exhibits about local history, and is related to the strike history because May Wing, who "lived the history," ran the Museum for some twenty years until shortly before she died in 1981 at age ninety. (See "History, Memory, and Commemoration," this volume).

Saint Victor's Church, (map location 1), 204 Portland Avenue. The walking tour stopped briefly at Saint Victor's, the church attended by many miners and their families. Father Edward Downey worked to maintain neutrality and fairness during the strike. During the union era he was known for his sensitivity to working-class issues, and is remembered particularly fondly for refusing to have a bell in his church, saying that he would not allow a bell to wake up men who had labored on night shift. The Goldfield Catholic Church is no longer standing, but Kathleen Welch Chapman reported that the priest at that church chose to be deported with the union men rather than denounce the union.[10]

Methodist Episcopal Church Site, (map location 7), 412 Victor Avenue and **the site of Rev. and Mrs. T. S. Lelands' House,** (map location 9), 550 South Fourth Street. The Church may have been the one served by Rev. T. S. Leland, a Methodist minister who supported the union during the strike and its aftermath. The Lelands' home was near the home of John Harper, Katherine Sturdevant's great-grandfather; the Lelands and Harpers were neighbors. We did not visit the church site, or the site of the Lelands' home, because the buildings no longer exist and because steep walking conditions made it a difficult hike for some members of our group. Instead, the walking tour stopped in front of the Baptist Church at the corner of Portland Avenue and North Fourth Street to discuss Rev. Leland and his role in the strike.

As mentioned in my article in this volume, "History, Memory, and Commemoration," on January 26, 1904, fifteen miners died on the Stratton's Independence Mine when the cable was drawn into the machinery and severed. Reverend Leland, pastor of the Victor Methodist Church, served on the coroner's jury that found management responsible for the fatal accident at Stratton's Independence, and he was one of the Ministerial Alliance members who established ties with the Cripple Creek and Victor Trades Assembly. Reverend Leland also belonged to a number of lodges in the district, including the Masons and Knights of Pythias, to which Adj. Gen. Sherman Bell also belonged. But the strike shattered the shared fraternity of local lodge halls.

Martial law was ended in the district July 26, and control returned to local authorities. Reverend Leland preached a sermon on July 31, 1904 that was sympathetic to the strikers and critical of the militia, the CA, and MOA. On August 10, the day after John Harper was beaten and driven from his home, three men threatened to hurt Leland if he did not leave town immediately. Leland instead began keeping armed men in his house for protection. Two union miners, Arthur Parker and L. R. Jenks, were there the night of August 28 when Sheriff Edward Bell and four deputies arrived, and exchanged shots. Bell said the miners tried to shoot him; Jenks and Parker said they thought it was an armed mob that had come to seize Reverend Leland, and that they opened the door as soon as they realized it was Bell. The sheriff arrested them and Leland and charged all three with assault to murder, conspiracy to murder, and assault with deadly weapons.

The charges against Parker, Jenks, and Reverend Leland of conspiracy to murder Sheriff Bell were eventually dropped. A jury in Castle Rock, on the plains south of Denver, acquitted Parker and Jenks of the remaining charges related to the raid on Leland's home. However, Leland's church required him to resign his Victor pulpit and accept reassignment away from the district. This may be why the Lelands' house was listed in a district directory soon after the strike as Mrs. Leland's home. She may have remained there for a time after her husband had to leave.[11]

Victor Record, (map location 6), 118 South Fourth Street. The union-owned Victor and Cripple Creek *Daily Press* closed just as the strike began, leaving the *Victor Record* the only pro-union newspaper in the district. The night of September 29, 1903, before Governor Peabody declared martial law in the district, soldiers, angry because the *Record* published official WFM statements and criticized the militia, arrested the *Record*'s workforce and took editor George Kyner and four printers to the military "bullpen." Emma F. Langdon, a linotype operator and member of the local Typographical Union, whose husband was one of the imprisoned workers, sneaked into the *Record* office and printed the paper. She delivered it to the bullpen brandishing the headline "Slightly Disfigured But Still in the Ring." For her actions, Langdon was the first woman elected an honorary member of the WFM.

At Peabody's instructions, the militia released the *Record* staff to Sheriff Robertson, charged them with criminal libel for reporting that a militia member was an ex-convict, and allowed their release after they posted bond. The night that Peabody declared martial law, Maj. H. A. Naylor informed Kyner that his paper was under military censorship, and that he could no longer publish official WFM statements. Neither could he print his editorial for the December 5, 1903 morning edition, which charged that Peabody had declared martial law because the troops had not broken the strike in "a manner satisfactory to the mine owners of the district." Kyner published his editorial, concluding, "The *Record* does not know at this time how far it will be allowed to express its opinion under the reign of military law, but we purpose [sic] to proceed just as if we lived in Free America."

After the Independence Depot explosion, George Kyner and the *Record* workforce were arrested yet again, held briefly, and released. On June 8, the *Record* urged the WFM to end the strike, not because it believed the union had anything to do with the explosion, but because it believed there would have been no violence if there had been no strike, and because it thought the strike was lost. That night eight armed men destroyed the office and machinery, an act for which the union was blamed despite the fact that the printers said that they recognized two

CA members in the wrecking party. A. C. Cole later said he was there, and knew that the CA was responsible. No one was arrested or charged with the destruction, but the *Record* resumed publication as an anti-union paper, leaving the unions without any public voice in the district. Peabody approved Kyner's claim for $4,206 in losses, and the state covered his losses.[12] The *Record* is also known locally as the paper where Lowell Thomas got his start as a journalist.

◊◊◊◊

The walking tour covered an area of only about four square blocks. Walking that small area emphasized spatially how linked the community institutions were, how so much could have transpired in one spot in the crucial hours before and after the Independence Depot explosion. It brought home as well that the community that was torn apart on June 6, 1904, was not an abstraction, but involved thousands of people who had daily walked the streets, worked the mines on too-aptly-named Battle Mountain, taken the F&CC train to the other district towns and mines, prayed in the churches, and come home to the houses from which some were then forcibly driven.

◊◊◊◊

Notes

Portions of this article come from my book, *All That Glitters: Class, Conflict, and Community in Cripple Creek.* It appears here with my permission and that of the University of Illinois Press. I am grateful to Chris Nicholl and Ed and Cherie Hunter, for helping to arrange the walking tour, to Joe and Marjie Stevens for allowing us inside the former Victor Miners' Union Hall, to Bridget Burke and Katherine Sturdevant for sharing their expertise at various sites, and to everyone who walked the tour on June 6, 2004.

1. The coroner's jury was impaneled by George A. Hall. Mr. Hall became coroner on June 6, 1904. At 9 A.M. James S. Murphy, manager of the Findley Mine, and Maj. H. A. Naylor of the state militia, called on the elected coroner, M. J. Doran, who had taken charge of the dead, and demanded that he deliver the bodies to a local undertaker, J. H. Hunt. Doran refused; Naylor told him that if he did not turn over the bodies, the militia would seize them. Doran complied to avoid further conflict, and was shortly deposed and replaced by Hall. On June 10 Hall's jury determined that the thirteen men whose deaths they were investigating died as a result of "an explosion of dynamite or other explosives" that were "exploded by an infernal machine" and that "said crimes are the result of a conspiracy entered into by certain members of the Western Federation of Miners and known, incited, and furthered by certain officers of that organization." See *Labor Disturbances in Colorado,* 247–48, 254. The objectivity of the investigation and the evidence on which the conclusion of conspiracy was based so soon after the explosion have been subjects of considerable debate.

2. It appears that Cole, whose politics were anti-union, was upset by what he believed his former allies had done to plan the violent events that ended the strike. Cole charged that prominent members of the MOA asked him to help with plans that were "similar to, or the same proposition, as the pulling off of the Independence station." When he refused, Cole said, he was "immediately relieved of his position as secretary of the Victor Citizens' Alliance." See Defense Attorney Richardson's notes, affidavits, and related material 1906–1909, State of Idaho vs. Steve Adams, Western Federation of Miners/International Union of Mine, Mill and Smelter Workers Collection, Western Historical Collections, University of Colorado, Boulder (hereafter cited as Richardson Notes), 72–76. Next to Cole's name someone, presumably Richardson, noted: "Pinkerton Operative No. 28." It should be noted that Cole was among those the union sued after the strike for assault to kill, malicious mischief, and criminal destruction of property, whether because of his actual behavior after the explosion or because it was unaware that he had been removed as

Citizens' Alliance secretary at the time is unclear. He did apparently, by his own account, participate in some of the anti-union actions in June 1904.

3. See *Labor Disturbances*, 249–50.

4. See Elizabeth Jameson, *All That Glitters: Class, Conflict and Community in Cripple Creek* (Urbana: 1998), 509–14.

5. Ibid., 76.

6. Ibid., 45, 65–66, 89, 102–03, 209. May Wing told me in 1976 that she thought the Woods family were victims, and that she had heard that they supported their union miners during the strike at great financial hardship to themselves. She was not in the district during the strike, and got much of her information from her aunt, Hannah Welch, whose husband John worked on the Gold Coin. I do not know whether this report is accurate. Interview with May Wing, Boulder, Colorado, March 6, 1976.

7. *Labor Disturbances*, 260; "The Situation in Colorado," *Miners' Magazine*, June 16, 1904, 7–8.

8. *Labor Disturbances*, 260, 267–68, 274–78, 295; George G. Suggs, Jr., *Colorado's War on Militant Unionism: James H. Peabody and the Western Federation of Miners* (Detroit: 1972), 112. See also "The Situation in Colorado," *Miners' Magazine*, June 16, 1904, 10; "The Situation in Colorado," *Miners' Magazine*, June 23, 1904, 10; "The Situation in Colorado," *Miners' Magazine*, July 7, 1904, 8.

9. I am grateful to the Victor Elks' Lodge for allowing us to tour their building. For information about the armory, see Cole account, Richardson Notes; *Labor Disturbances*, 247–49; Suggs, *Colorado's War*, 110–11. Besides Robertson, the deposed officials included Coroner M. J. Doran; Cripple Creek Marshal W. J. Graham; Cripple Creek Night Marshal Fred Harding; Cripple Creek Justices of the Peace J. P. Thomas and David Kelley; Victor Aldermen J. W. Murphy and J. J. Tobin; Victor Police Judge Michael Gibbons; Victor Night Marshal Michael Lamb; Victor Jailer James Printy; Independence Marshal Harvey Starbuck; Goldfield Marshal J. J. Brothers; Goldfield Night Marshal Robert McCarthy; Anaconda Marshal E. E. Scott; several members of the Anaconda City Council; Goldfield Aldermen J. F. Daugherty, Arthur Childers, Christopher Miller, M. D. Morrison, and A. J. Burke; the Goldfield city clerk, treasurer, street commissioner, and marshal; and Deputy District Attorney J. C. Cole. *Labor Disturbances*, 251–52. Edward Bell and Sherman Bell were not related.

10. Interviews with Kathleen Chapman, Wheat Ridge, Colorado, April 27, 1979.

11. *Labor Disturbances*, 307-08, 310-14; *Miners' Magazine*, November 24, 1904, 4; *Miners' Magazine*, December 29, 1904, 3; "The Situation in Colorado," *Miners' Magazine*, January 5, 1905, 10.

12. "The Cripple Creek Situation," *Miners' Magazine*, October 8, 1903, 6; Labor Disturbances, 209-10, 250-51, 261-64.

Mining & Labor History
A Selected Bibliography Available at the Pikes Peak Library District's Special Collections

Beshoar, Barron B. *Out of the Depths: The Story of John R. Lawson, a labor leader.* Denver: The Colorado Labor Historical Committee of the Denver Trades and Labor Assembly, 1957.

Brundage, David Thomas. *The Making of Western Labor Radicalism: Denver's Organized Workers, 1878–1905.* Urbana: University of Illinois Press, 1994.

Colorado Mine Operators' Association. *Criminal Record of the Western Federation of Miners, From Coeur d'Alene to Cripple Creek, 1894–1904.* Colorado Springs, Colo.: The Association , 1904.

Haywood, Big Bill. *Bill Haywood's Book: The Autobiography of William D. Haywood,* New York: International Publishers, 1929.

Jameson, Elizabeth. *All That Glitters: Class, Conflict, and Community in Cripple Creek,* Urbana: University of Illinois Press, 1998.

Jones, Mother. *Autobiography of Mother Jones,* New York: Arno, 1969.

Kimmett, Leo. *Twelve Thousand Cripple Creek Miners,* Cañon City, Colo.: L. Kimmett, 1991.

Knight, Harold V. *Working in Colorado: A Brief History of the Colorado Labor Movement,* Boulder: University of Colorado, Center for Labor Education and Research, 1971.

The Labor News [microform] 1904–1917.

Langdon, Emma F. *The Cripple Creek Strike: A History of Industrial Wars in Colorado, 1903–4–5: Being a Complete and Concise History of the Efforts of Organized Capital to Crush Unionism,* Denver, Colo.: Great Western Publishing, 1904–05.

Levine, Brian H. *Cripple Creek Gold: A Centennial History of the Cripple Creek District,* Lake Grove, Ore.: The Depot, 1988.

Levine, Brian H. *The Portland, Colorado's Richest Gold Mine,* Victor, Colo.: Syzygy Gold Mining Co., 1989.

Lingenfelter, Richard E. *The Hardrock Miners; A History of the Mining Labor Movement in the American West, 1863–1893*, Berkeley: University of California Press, 1974.

Long, Priscilla. *Where the Sun Never Shines: A History of America's Bloody Coal Industry*, New York : Paragon House, 1989.

Lovin, Hugh T. *Labor in the West*, Manhattan, Kan.: Sunflower University Press, 1986.

Lynde, Francis. *Cripple Creek 1900: A Colorado Mining Camp, an Eyewitness Account*, Olympic Valley, Cal.: Outbooks, 1976.

MacKell, Jan. *Cripple Creek District: Last of Colorado's Gold Booms*, Charleston, SC: Arcadia Pub., 2003.

Mazzulla, Fred. *The First 100 Years: A Photographic Account of the Rip-roaring, Gold-plated Days in World's Greatest Gold Camp, Including Views of the Best Fires, Labor Wars, Murders, Parlor Houses and Politicians in Cripple Creek and the Pikes Peak Region*, Denver, Colo.: A. B. Hirschfeld Press, 1964.

The Miners' Magazine [microform] 1900–1905.

Neuschatz, Michael. *The Golden Sword: The Coming of Capitalism to the Colorado Mining Frontier*, New York: Greenwood Press, 1986.

Rastall, Benjamin McKie. *The Labor History of the Cripple Creek District*, Madison, Wis., 1908.

Sargent & Rohrabacher. *The Fortunes of a Decade: A Graphic Recital of the Struggles of the Early Days of Cripple Creek, the Greatest Gold Camp on Earth*, Colorado Springs, Colo.: Old Times Publishers Co., 1979.

Spell, Leslie Doyle. *Forgotten Men of Cripple Creek*, Denver, Colo.: Big Mountain Press, 1959.

Sprague, Marshall. *Money Mountain: The Story of Cripple Creek Gold*, Boston: Little, Brown, 1953.

Taylor, Robert Guilford. *Cripple Creek Mining District*, Palmer Lake, Colo.: Filter Press, 1973.

Selected Archival Collections of Interest

Empire State Consolidated Gold Mining Company, Ledger, MSS 0061.

Irving Howbert, Papers, MSS 0168.

Mines and Mining Documents, Collection, MSS 0260.
John T. Hannah Mining Stock Certificate Collection, Mining stock certificates from gold mines in Victor/Cripple Creek Ca. 1895–1917.

Selected Genealogy Resources of Interest

Kimmett, Leo. *Twelve Thousand Cripple Creek Miners,* Cañon City, Colo.: L. Kimmett, 1991.

Sturdevant, Katherine Scott. *Bringing Your Family History to Life Through Social History,* Cincinnati, Ohio: Betterway Books, 2000.

Sturdevant, Katherine Scott. *Organizing and Preserving Your Heirloom Documents,* Cincinnati, Ohio: Betterway Books, 2002.

Map, Cripple Creek Gold District, 1896. *From Special Collections, Pikes Peak Library District.*

SYMPOSIUM CONTRIBUTORS

Keynote Address

Elizabeth Jameson, Ph.D., is a history professor and holds the Imperial Oil-Lincoln McKay Chair in American Studies at the University of Calgary. She is the author of *All That Glitters: Class, Conflict, and Community in Cripple Creek* (1998); *Building Colorado: The United Brotherhood of Carpenters and Joiners of America in the Centennial State* (1984); and the co-editor of *Writing the Range, Race, Class, and Culture in the Women's West* (1997) and *The Women's West* (1987).

Panel: Women Voters, Writers, and Workers in the Cripple Creek Strike

Bridget Burke is the Director of the American Alpine Club Library in Golden, Colorado. She has a Masters Degree in Library and Information Science from the University of Wisconsin, and a Masters Degree in American History from Wesleyan University. She has worked in research libraries for fifteen years and has held curatorial positions at the Colorado Historical Society and the Yale Collection of Western Americana. She is currently working on a history of female strikebreakers in the printing trades in the 1950s.

Marcia Tremmel Goldstein is a Ph.D. Candidate, University of Colorado at Boulder, History Department, who specializes in the history of women and labor in the American West. She has been a history instructor at University of Colorado at Colorado Springs, University of Colorado at Boulder, and Metropolitan State College of Denver. She is the author of *Denver Women in Their Places: A Guide to Women's History Sites* (2002).

Jan MacKell is a historian and archivist at the Cripple Creek District Museum in Cripple Creek, Colorado, where she served on the Historic Preservation Commission for four years. She is the author of *Cripple Creek District: Last of Colorado's Gold Booms* (2003). Her second book, *Brothels, Bordellos & Bad*

Girls: Prostitution in Colorado, 1860–1930 was published by the University of New Mexico Press in 2004. She has written over 700 articles about Colorado history, has worked on historical video-documentaries about Cripple Creek, and is currently working on several other book projects.

Panel: Militia and Mine Owners

Ed Hunter is a mining engineer and past president of the Mining History Association. He is a local historian and works on historic preservation of trails and mining sites in the Cripple Creek Mining District. He is the co-author of *A Concise History of Mine Hoisting: From its Earliest Beginnings Through Winfield Scott Stratton's Independence Hoist* (2002).

Allan C. Lewis is a freelance writer and researcher of western American railroad history. He is the author of *Florence & Cripple Creek Railroad, Forty Miles to Fortune* (2002), and he recently published *Railroads of the Pike's Peak Region, 1870–1900* (2004) and *Colorado & Southern Railway: Clear Creek Narrow Gauge* (2004). His presentation at the symposium, "The Colorado National Guard: The Other Side of the Story," was not available for this publication.

Concluding Presentations

Katherine Scott Sturdevant, great-granddaughter of miners' union leader John Harper, is Chair of History at Pikes Peak Community College in Colorado Springs, where she has taught history since 1986. As a teacher and speaker, Kathy's specializations have included American social, Colorado, western, women's, Pikes Peak regional, and family history. She has published two books: *Bringing Your Family History to Life through Social History* (2000) and *Organizing and Preserving Your Heirloom Documents* (2002). She is the recipient of local, state, and national teaching awards and a regular participant in local history programs.

Richard Marold served as the Director of the Cheyenne Mountain Heritage Center and as the editor of the award winning quarterly journal, *Kiva: The Journal of the Cheyenne Mountain Heritage Center*. He portrays historical characters including President Franklin D. Roosevelt, local educator Lloyd Shaw, and Winfield Scott Stratton, the first multimillionaire from the Cripple Creek District. He has also been featured on The Thomas Jefferson Hour® and teaches history at Colorado State University in Pueblo. He was in character at the symposium as Winfield Scott Stratton.

Richard Marold as Winfield Scott Stratton. *Photograph courtesy of Cara Newman, Howling Wolf Photography.*

About Pikes Peak Library District

Pikes Peak Library District (PPLD) is a nationally recognized system of public libraries serving a population of more than 500,000 in El Paso County, Colorado. With 12 facilities and two bookmobiles, PPLD responds to the unique needs of individual neighborhoods and the community-at-large. PPLD has an employee base of 375 full and part-time staff, and utilizes roughly 1,500 volunteers. It strives to reach all members of the community, providing free and equitable access to information and an avenue for personal and community enrichment.

PPLD is rated 7th among library systems its size in the country. Volume of circulations, number of visits, and hours of access contribute to the ranking. PPLD is also recognized for its commitment to diversity, its quality programming, and its excellent customer service.

Forthcoming Regional History Series Publications

Legends, Labors & Loves
William Jackson Palmer, 1836–1909

Zebulon Montgomery Pike & His 1806–1807 Expedition
"To Spare No Pains"
A Bicentennial Commemoration

INDEX

Acton, Joe, photo of, 18
Adams, Alva, 46–47
Altman WFM local. *See* Free Coinage Miners Union No. 19
American flag: desecration of, 66
Anaconda, Colo., photo of, 5
Anthony, Leonel "Nell" Ross Campbell, 54–58, 68; on martial law, 67; photo of, 55; Polly Pry, 55
Arapahoe County, Colo., 39
Arizona, 84, see also Bisbee, Globe
Arizona Territorial Industrial School, 96
Aspen, Colo., 4, 37
Augustine, Gus, 102

Baltimore Hotel, 103
Battle Mountain, 3, 107
Beck, Melvin, killed, 9
Bell, Edward, 103, 110, 112
Bell, Sherman, 4, 10, 11–12, 44–45, 62, 66–67, 108–109, 112; fan of *Polly Pry*, 58, 61; guarantees John Harper protection, 94; poem mocking, 57
Bellism, 66. *See also*, Bell, Sherman
Bennett Avenue (Cripple Creek), photo of, 62
Bisbee, Ariz., 85, 95; deportations, 26
Blacklisting, 23, 83, 85, 95
Boise, Idaho, 96
Boycotts, 18–19
Bradford, Mary C. C., 46; photo of, 46
Bryant, Frederick Bridgeman (Cpl.), photo of, 8
Bulger, Eddie, harassed, 21
Bullpen, 20, 34, 44–45, 59, 70, 92, 110, photo of, 12. *See also* National Guard Armory
Bull Hill: strike, 7, 22
Burke, Bridget, 115, 121; "Is Colorado in America?: Emma Langdon, Polly Pry and the Colorado Labor Wars," 53–70; photo of, 102
Burns, James, 3, 7, 10, 78
Burns, Robert, poetry mentioned, 86, 94

Calaconda Mine: photo of workers, 18
Caldwell, Idaho, 11
Cameron, Colo., 16
Campbell, Slim, murders prostitute, 75
Cañon City, Colo., 4, 93-94
Carley, John, 109
Carlton, Albert, 3, 10, 101, 108
Castle Rock, Colo., 112
Chapman, Kathleen Welch, 4, 7, 15–17, 19–22, 27, 111; and deportees, 20; recalling strike defeat, 23
Churchill, Caroline Nichols, 38
Citizens' Alliance, 9–10, 26, 90, 93, 103–104, 108, 110, 112, 114, 116; deposed officials listed, 116; forced resignations, 110
Clear Creek County, Colo., 8, 9
Coal fields: accident in, 7; strikes in, 8–9, 43, 55, 58
Coeur d'Alenes, Idaho, 11, 13, 25
Cole, A. C., 103, 109, 114–115
Colonial Dames, 65
Colorado City, Colo., 43

Colorado City Mill & Smeltermen's Union No. 125, 7
Colorado Fuel & Iron Company, 45, 58
Colorado Labor Wars, 25–26, 56, 67; provoked, 43; strikes, 1902–1903, 43
Colorado Midland Terminal, photo of, 62
Colorado Mine Owners' Association, 13
Colorado School of Mines, 88
Colorado Springs, Colo., 7, 37, 98-99
Colorado Springs & Cripple Creek District Railroad, 107
Colorado State Federation of Labor, 44; women's auxiliary, 54, 60
Colorado State Federation of Women's Clubs, 41
Colorado Trading & Transfer Company, 108
Conine, Martha, 40–41, 46; photo of, 46
Conrad (pressman), 60
Constitutional rights, 25, 67; martial law ended, 112; violations of, 10, 44, 63
Cooperative stores (WFM), 10, 90-94
Crapper Jack's (saloon), photo of, 76
Cressingham, Clara, 40
Cripple Creek, Colo., 3, 5–6, 53; homes deserted, 23-24; mining district, map of, 100, 120
Cripple Creek Morning Journal (newspaper), 74
Cripple Creek strikes, 3, 26, 39; of 1894, 4–5; 1904 commemoration, 3, 24–27, 53, 101–116; defeat recalled, 23; disputes and minor conflicts, 18; military occupation, 6; of 1903–1904, 6; political impact of, 45; prostitutes support of, 75; Waite Agreement, 38
Cripple Creek and Victor Trades Assembly, 112

Daily Press (newspaper), See Victor and Cripple Creek Daily Press
Darrow, Clarence, 32, 96
Daughters of the American Revolution, 63, 65
Davis, John, 105
Deadwood Mine, 101
Democratic Party, 8, 43, 46–47, 55, 61
Denver, Colo., 11, 36, 37, 40-43, 46-47, 56, 61, 65, 93-94
Denver Post (newspaper), 55–56
Denver Typographical Union, 55, 60
Deportation of union members and sympathizers, 6, 9, 10, 19, 26, 46, 61, 83, 85, 92, 95, 108; photo of, 22. *See also* Miners: deportation of
Depression, economic: in Colo., 37
Dickey, Mary Frances "Fannie" Withers, 89
Dickson, George E., 32
District Trades Assembly, 54
Doran, Catharine, 14, 20
Doran, Edward, 14–15, 20–21
Doran, M. J., 103, 115
Doran, Mary, 15, 20
Downey, Edward (Fr.), 111
Doyle, Jimmy, 3, 78
Dunnville, Colo., 109

Economic Gold Extraction Company, 107
Election fraud, 47
Electoral reform, 48
Elks' Lodge, 109
El Paso County, Colo., 8
El Paso Mine: Calaverite crystals, 77; headframe and sorting house, photo of, 79; Jewelry Stope, photo of, 78
Ensley, Elizabeth, 37
Evens, Laura, 74
Excelsior Engineer's Hall, 60

Findley Mine, 101; non-union miners killed, 10
Flag poster (WFM) and flag desecration lawsuit, 56, 62-70
Florence & Cripple Creek Railroad, 9, 101, 109; depot in Victor, 108
Fourth of July, 15
Fourth Street, Victor, 74
Free Coinage Miners Union No. 19: and 1894 strike, 38; members accused of bombing, 105-106
Fremont County, Colo., 39
French, F. D., 104

Gatheridge, Charlie, 4
Gaughan, James, 110
Gibson's Camp, Ariz., 95
Gillette, Colo., 74, 100
Globe, Ariz., 95-96
Golden Crescent Water & Light Company, 107
Goldfield, Colo., 15, 17, 23, 70; photo of, 8, 20, 21
Goldfield, Nev., 96
Goldfield Catholic Church, 111
Goldstein, Jeff, photo of, 102
Goldstein, Marcia Tremmel, 121
"Meet Me at the Ballot Box: Women Voters, Striking Miners, and Colorado Labor Politics, 1893–1904," 35–52; photo of, 102
Gold Coin Club, 17–18, 107; photo of, 17
Gold Coin Mine, 17, 19, 88, 107; union cards checked, 19
Gold ore values, 77
Golden Cycle Mine, 18
Gonzales, Delfido, photo of, 34

Haag, Henry, 102
Habeas corpus, 66; rights suspended, 9
Hagerman, James J., 50
Hall, George A., 115
Halthusen, Sally, 73
Hamlin, Clarence, 105, 108
Harman, Thomas H., 32
Harnan, John, 78
Harper, Archibald Leslie "Archie," 84–86, 88, 95; arrested, 94; deported, 94; at law school, 94; moved to Vallejo, Calif., 97; opened a law practice in Globe, Ariz., 96; photo of, 94, 97; represented "Big Bill" Haywood at Idaho trial, 96; taken from home and released, 93
Harper, Barbara Leslie, 87
Harper, Barbara Pearl (Scott), 83-84, 97
Harper, Grace, 88, 93
Harper, Herbert John, 88
Harper, John, 83–99, 111; charitable donations described, 86; deported, 83, 85, 91, 95; disavowed violence, 90; investments, 86; life of, 87; manager of the Victor Cooperative Store, 90; photo of, 84; physically

described, 84; sweated, 92; taken from home and beaten, 92
Harper, John Herbert, Jr., 85, 88

Harper, John Leonard "Len," 96-97
Harper, Kate (Dickey), 85–86, 89, 97
Harper, Katherine Margam, 87–89, 95; photo of, 89
Harper, Kathleen (Puckett), 87
Harper, Leslie Archibald, 97
Harper, Marian Katherine, 97
Harper, Pearl, 88, 93
Haywood, William "Big Bill," 4, 11, 43, 63, 70, 96; and flag desecration suit, 63, 66; on John Harper being beaten and robbed, 93–94
Hazlett, Ida Crouch, 60
High-grade gold ore, 77–82; sale of, 79
High-grading, 26, 77–82; photo of lunch bucket used for, 81; prevention of, 80–82; use of candleholder for, 78; use of lunch bucket for, 78
"High-Grading: Theft or Wage Supplement Right?," by Ed Hunter, 77–82
"History, Memory, and Commemoration: The Cripple Creek Strike Remembered," by Elizabeth Jameson, 3–33
Holly, Carrie, 40
Huerfano County, Colo., 39, 47
Hunt, J. H., 103, 115
Hunter, Cherie, 115
Hunter, Ed, 115, 122; "High-Grading: Theft or Wage Supplement Right?," 77–82; photo of, 102
Idaho, 38. *See also* Caldwell, Coeur d'Alene
Independence, Colo., 3, 75; railroad depot explosion, 6, 10–11, 26–27, 61–62, 101
coroner's jury, 115; investigation, 109; mine owners responsible, 26; photo of, 25; union responsible, 14; vigil, 101–102; witness killed, 11
Independence Mine, 26, 112; hoist accident, 9–10; ore values, 78; photo of hoist, 10
Industrial Workers of the World, 5, 55
International Harvester (Co.), 96
International Union of Mine, Mill & Smelter Workers, 106. *See also* Western Federation of Miners
Ireland and the Irish, 14
"Is Colorado in America?" (poster), 44, 63–65; flag desecration suit, 63–65; photo of, 64, 70
"Is Colorado in America?: Emma Langdon, Polly Pry and the Colorado Labor Wars," by Bridget Burke, 53–70

Jameson, Elizabeth, 1, 88, 89, 98, 121; cited, 83, 88; "History, Memory, and Commemoration: The Cripple Creek Strike Remembered," 3–33; "Talking the Walk: The 1903–1904 Strike Centennial in the Cripple Creek District," 101–116; photo of, 102
Jenks, L. R., 112
"John Harper, Deported Miner: The Cripple Creek Strike as Family History," by Katherine Scott Sturdevant, 83–99

Johnson, Jack, 107
Jones, Mary Harris "Mother," 44–45, 61; photo of, 44

Kansas, 6, 10, 66, 85, 92, 108
Kelly, Jerry, photo of, 34
Kennedy, Mrs., 47
Kerr, Samuel, 87
Klock, Frances, 40
Knights of Labor, 14, 37
Knights of Pythias, 112
Knights of St. John, 16
Krotz, L. E., photo of, 34
Kyner, George, 112–113; photo of, 12

Labor Day, 15, 16, 17; photo of parade, 16
Langdon, Charles G., 54; imprisoned in bullpen, 59; photo of, 12
Langdon, Emma F., 12, 13, 47, 54–56, 98; on the 1903–1904 strike, 35, 45, 54, 61, 67, 68; first female member of WFM, 43; and the "Is Colorado in America?" poster, 65; photo of, 54; and *The Victor Record,* 43, 59–61, 113
Langdon, Frederick, 54; photo of, 12
Langdon, Lucile, 55
Laundry girls law, 41
Leadville, Colo., 4, 7, 15, 21, 86, 88
Legislature, Colo. State: women elected to, 40–41
Leland, T. S. (Rev.), 93, 111–112
Leo the Lion, 3
Lewis, Allan C., 122
Lindsey, Ben (judge), 96
Livingston, Lola, 73
Ludlow, Colo., 25
Ludlow Memorial: vandalized, 6, 14, 27
Lump Gulch, 12, 24
Lunch bucket: photo of, 81; used for high-grading, 78
Lynching, 9, 11, 92-93, 110; photo of, 11
MacKell, Jan, 121; "Prostitution During the Cripple Creek District Labor Wars, 1904", 71–76
MacLaren, Ian, books, 86, 97
MacNeill, Charles, 3, 7-8
Maki, Henry, 65, 70; photo of, 64
Mannix, Frank, 60
Marold, Richard, 123
Mary McKinney Mine, photo of, 5
Masonic Lodge, 97, 112
McClelland, Thomas, 10
McCluskey, Belle, 3
McConaghy, James, 15
McConaghy, John, 15
McConaghy, Kate Doran, 15
McCormick, Charles, killed, 9
McCoy, Herbert, 102
McDonald, Jesse (gov. of Colo.), 48
McGee, Roscoe, 105
McGorraity, Mamie, 75
McGuirk, Michael, 32
McIntire, James (gov. of Colo.), 40
Mee, Pat, 32
"Meet Me at the Ballot Box:" Women Voters, Striking Miners, and Colorado Labor Politics, 1893–1904," by Marcia Tremmel Goldstein, 35–52
Merchants Biscuit Company, 41; photo of, 42
Meredith, Ellis, 36, 46; photo of, 46
Methodist Episcopal Church, 111

Mexican-Americans, 96
Militia, 10
Miller, A. Q., 60
Miller, Alfred, 105
Miller, Christopher, 105
Miller, Haillie, 71
Miners 33–34; deportation of, 6, 10, 26, 43, 108–109; escorted by National Guard, photo of, 22; families harassed, 21; at Independence railroad depot, 102; killed, 10; "mug shots," 95; non-union, 63; threatened, 47; photo of, 33; *Polly Pry* cartoon described, 66; and the Populist Party, 37; use of violence examined, 14, 27
Miners' Magazine, 60, 90–91, 92
Mine owners: blamed for deaths, 13; and local government, 13, 29; and the Republican Party, 39, 43
Mine Owners' Association, 11, 13, 22, 26, 27, 44, 60, 90, 108, 110; deposed officials listed, 116; forced resignations, 110; intimidation, 11
Minister, Mrs. Fred, 46-47
Ministerial Alliance, 111-112
Mitchell, John, 58
Moffat, David H., 50
Moffat mines, 10
Moore, Harry, 109
Morning Glory Mine, 5
Moyer, Charles, 11, 43, 63, 70; arrested, 44; flag desecration suit, 63–66
Mt. Pisgah (Cripple Creek), 4
Muhliesen, Arthur, 102
Mullaney, Pat, photo of, 34
Murphy, James S., 115
Murrah, John, photo of, 18
Myers Avenue (Cripple Creek), 3, 74

National Guard, 8–9, 59, 62; Company L, 9, 103, 105, 109; dispatched by Gov. Peabody, 9; instructed to clean out WFM, 103; photo of, 8, 11, 12, 20, 21, 22
National Guard Armory, Victor, 105, 108, 109
Naylor, H. A., 104, 109, 113, 115
Nevada. *See* Tonopah
New Mexico, 6, 10, 92, 108. *See also* Silver City
Newspapers, 5; freedom of the press suspended, 9, 59–60, 113; publicizing boycotts, 18; reporting on strikes, 58, 67. See also *Cripple Creek Morning Journal; Denver Post; Polly Pry; Queen Bee, The; Rocky Mountain News; Victor and Cripple Creek Daily Press*
Nicholl, Chris, 28, 95, 115, photo of, 102
Nicholl, Nick, photo of, 102
Non-Partisan Colorado Equal Suffrage Association, 37

O'Connell, Michael, 104
Oral tradition and oral history, 15, 83–99
Orchard, Harry, 11, 12, 26, 75, 82

Packer, Alferd, 55
Parfet ,T. H., 90, 92
Parker, Arthur, 112
Parsons, L. F., 110
Peabody, James Hamilton (gov. of Colo.), 4, 8–9, 35, 43, 44, 45, 47, 54, 57, 67, 70; contests Adams's election, 47; declared

martial law, 112
Peabodyism, 46
Penrose, Spencer, 3, 7-8, 73–74
People's Party. *See* Populist Party
Pettibone, George, 11
Pike's Peak Power Company, 107
Pinkerton Detective Agency, 8; business cards used in flag desecration defense, 66; and Harry Orchard, 26; infiltrates union, 8
Pinnacle Park, 15, 16
Polly Pry (newspaper), 55, 56–59, 61–62, 66; and Denver society, 57; and flag desecration suit, 63–64; photo of, 57
Populist Party, 4, 8, 14, 37, 39–41, 43
Porter, Rufus L., 82
Portland Mine, 8, 10, 110
"Prostitution During the Cripple Creek District Labor Wars, 1904," by Jan MacKell, 71–76
Prostitution and prostitutes, 3–4, 71–76; customers, 72; investing in mines, 72; as a means of spying, 71; photo of, 72; regulation of, 74
Pryor, Beulah, 4
Pueblo, Colo., 37, 40

Queen Bee, The (newspaper), cited, 38

Rastall, Benjamin McKie, cited, 13–14
Republican Party, 39, 40, 103; leaders provoked the labor wars of 1902–1904, 43; makes accusations, 47; women in the Colo. State House of Representatives, 40
Reynolds, Minnie, 36, 39
Richmond, H. J., photo of, 12
Rickard, Thomas A., 77
Robertson, Henry, 103–104, 109–110, 113; photo of [?], 11
Robinson, Harry, photo of, 34
Rocky Mountain News (newspaper), 55
Root, Candace, 73
Ross, Edward, 102

Saint Victor's Church, 111
Salida, Colo., 74
Scotch-Irish, 89
Scotland and the Scottish, 86-99
Shetland Islands, Scotland, 87, 99
Shafroth, John Franklin (gov. of Colo.), 48
Shanklin, William, 102
Sherman Silver Purchase Act: rescinded, 37
Silver and Gold, The (magazine), 94
Silver City, N. Mex., 14
Silver currency standard, 37
Social life, 15–17
Socialist Party, 5, 55
Sons of the American Revolution, 63
Speer, Robert, 47
Stahl, J. D., photo of, 34
Standard Mine, 8
Steunenberg, Frank, (gov. of Idaho), 11
Stevens, Joe, 106, 115
Stevens, Marjie, 106, 115
Stewart, Thomas, beaten, 9
Stratton, Winfield Scott, 3, 10, 73; photo of, 73
Sturdevant, Katherine Scott, 97–99, 115, 122; experienced retaliation as union leader, 91; "John Harper, Deported Miner: The Cripple Creek Strike as

Family History," 83–99; photo of, 97, 102
Sturdevant, Rick, 91, 98–99; photo of, 102
Suffrage rights: and "Mother" Jones, 45; measure passed in Rocky Mountain states, 38; for women, 35–37
Sullivan, J. C., 44
Sweating, 61; of John Harper, 92
Sweet, W. L., photo of, 12

"Talking the Walk: The 1903–1904 Strike Centennial in the Cripple Creek District," by Elizabeth Jameson, 101–116
Teller County, Colo., 39, 47, 53, 60, 62, 66
Teller County Mining Supply Company, 107
Telluride, Colo., 8, 9, 43, 47, 58, 63, 64, 65, 70
Third Street, Victor, 74
Thomas, Lowell, 114; defends prostitutes, 75; Museum, 111
Tincup, Colo., 14, 24
Tonopah, Nev., 14
Topic Theater, 74
Travell, W. W., 105
Trilby Mine, 22
Trinidad, Colo., 58
Tutt, Charles, 3, 7-8, 73–74

Unions and union labor, 5, 8; benefits, 15; blacklisting, 23; boycotts, 18–19; Denver "food strike," 58; as reported in *Polly Pry*, 58; wages and hours, 8–9, 18, 38, 42, 45, 48, 50, 77. *See also* Miners
United Mines Transportation Company, 107
United Mine Workers of America, 43–44, 58
University of Colorado, Boulder, Law School, 88, 93, 94
Utah, 38, 42

Verdeckberg, Edward (Col.), 45
Vernon, Hazel, 3
Victor, Colo., 3, 88; brothels in, 74; fire of 1899, 74; map of, 104; Museum, 4; walking tour, 103–114
Victor Armory, 105, 109
Victor Avenue, 74-75
Victor and Cripple Creek Daily Press, 5, 18, 112
Victor Company, 45
Victor Cooperative Store, 90; raided, 92
Victor Elks' Lodge, 109
Victor Fuel (Co.), 58
Victor Miners' Union Hall, 105–106; battle at, 110; photo of, 24
Victor Record (newspaper), 12, 54, 56, 59, 62, 113-114; urges WFM to end strike, 113
Vigil, 100-102
Victor Miners Local No. 32, 24, 90-99, 103, 105; photo of parade, 16
Vindicator Mine: explosion at, 9–10, 12, 26, 82
Violence, 10, 12-13, 26, 34, 47, 75, 91-95, 98, 101-114

Waite, Celia (first lady of Colo.), 39
Waite, Davis H. (gov. of Colo.), 4, 37; and 1894 strike, 38; photo of, 39
Waite Agreement, 38, 43, 49–50
Wales and the Welsh, 14, 87-89
Walking tour. *See* Jameson
Welch, Annie, 15

Welch, Edward, 15
Welch, Hannah Doran, 7, 14, 15, 20–21, 26–27
Welch, John, 7, 14–16, 17–19, 27; dies, 23; injured, 23; slated for deportation, 19; son John, 15
Welch, Tom, 7, 22; harassed, 20–22
Western Federation of Miners, 4–5, 7–9, 13, 14, 22, 38, 94; and 1894 strike, 38; charged with crimes, 13; and cooperative stores, 90; and Independence Depot bombing, 110, 115; leaders implicated as criminals, 11, 61; leaders implicated in murder, 105; officials described, 58; poster with American flag, 44; and social pressure, 19; urged to end strike, 113; Women's Auxiliary, 16, 26, 41, 43. *See also* Victor Miners Local No. 32
Western Steam Laundry, 41
Whitecapping, 10, 61
Whitecaps, 93
Wilkinson, Leslie, 4
Wilkinson, Ruby, 4
Wing, May McConaghy, 4–6, 15, 20, 27, 111
Womack, Bob, 3
Women, 3–5, 15, 16, 35-69, 71-76, 88-89, 93; and the vote, 36, 46, elected to Colo. legislature, 40–41; and tradition, 89; wages and hours, 40, 48; working conditions, 41. *See also* Suffrage rights
Women's Press Club, 56
Women's Trade Union Unity League, 41
Wood, Myron: photo by, 17
Woods, Frank, 107
Woods, Harry, 107
Woods, Warren, 107
Woods family, 19, 116
Woods Investment Company, 16–17, 107; succumbed to MOA pressure, 107
Wyoming, 38